W9-DIW-092

Introduction to Freshwater Vegetation

Introduction to Freshwater Vegetation

DONALD N. RIEMER

Department of Soils and Crops
Rutgers University
New Brunswick, New Jersey

AN avi BOOK
Published by Van Nostrand Reinhold Company
New York

An AVI Book
(AVI is an imprint of Van Nostrand Reinhold Company Inc.)
Copyright © 1984 by Van Nostrand Reinhold Company Inc.

Library of Congress Catalog Card Number 83-22477

ISBN 0-87055-448-4

Printed in the United States of America

Van Nostrand Reinhold Company Inc.
115 Fifth Avenue
New York, New York 10003

Van Nostrand Reinhold Company Limited
Molly Millars Lane
Wokingham, Berkshire RG11 2PY, England

Van Nostrand Reinhold
480 La Trobe Street
Melbourne, Victoria 3000, Australia

Macmillan of Canada
Division of Canada Publishing Corporation
164 Commander Boulevard
Agincourt, Ontario M1S 3C7, Canada

16 15 14 13 12 11 10 9 8 7 6 5 4 3 2

Library of Congress Cataloging in Publication Data

Riemer, Donald N.
    Introduction to freshwater vegetation.

    Includes bibliographies and index.
    1. Freshwater flora. 2. Aquatic weeds—Control.
3. Freshwater ecology. I. Title.
QK932.R54 1984   581.5'2632   83-22477
ISBN 0-87055-448-4

Contents

Preface

There are several excellent reference books available on vascular aquatic plants. Among these are Sculthorpe's *The Biology of Aquatic Vascular Plants;* Hutchinson's *A Treatise on Limnology, Vol. III, Limnological Botany,* and the older but still valuable book by Arber entitled *Water Plants.* Despite their value as reference books for specialists in the field, they are not well suited as introductory texts because they require considerable background information and specialized vocabulary on the part of the reader.

This book is intended to fill the need for such an introductory text. It is written primarily for students in ecology, botany, wildlife management, fisheries management, and related fields, who need a college-level introduction to vascular plants of aquatic habitats. It assumes that the student has had an introductory course in inorganic chemistry and an introductory course in biology, but no other background is assumed.

A second, but no less important, use for the book is in the non-degree courses now offered by many institutions. For example, there are numerous training courses now available for persons who wish to be licensed as certified aquatic pesticide applicators. Colleges also offer non-degree courses for park superintendents, golf course superintendents, and other professional people whose duties include the management of water. All will benefit from this book.

Part I is a brief introduction to aquatic environments. It is not intended to replace a course in limnology but gives the student sufficient background to understand the chapters that follow.

Part II focuses on the biological aspects of the vascular plants of fresh waters. The emphasis is on the differences between terrestrial and aquatic species, assuming that the student has had a course in biology and understands such basics as the importance of photosynthesis and has at least a cursory knowledge of the gross anatomy and morphology of plants.

Part III deals with the interactions between aquatic plants and mankind. This section of the book is more practical and applied than the previous two sections and considers aquatic plants as pests, as assets, and to a very limited degree as crops to be established and managed as are terrestrial crops.

Thus, the book has great breadth and covers almost every facet of the higher vegetation of fresh waters. It will be useful to a variety of students in many different learning situations.

Donald N. Riemer

Acknowledgments

Very special thanks are due to my wife, Janet Tepper Riemer, for her continued encouragement and understanding throughout the preparation of this book. She spent many hours proof reading the manuscript, preparing illustrative materials, and making dozens of useful suggestions.

Grateful appreciation is also expressed to Mr. David W. Platt of Rutgers University for his technical assistance with the word processing system on which the manuscript was written. Without his patient, cheerful, expert advice the word processor could not have been employed. Finally, thanks are due to to Dr. Harry L. Motto of Rutgers University for reviewing Chapter 5 and offering valuable criticism.

Part I

An Introduction to
Aquatic Environments

1

Diversity of Aquatic Environments

There are numerous references in books and other writings to the "aquatic environment." In one sense, there is no such thing as *the* aquatic environment. Aquatic environments vary among themselves just as greatly as terrestrial environments do. The differences between deserts and rain forests, mountain tops and prairies, and temperate woodlands and tundras are obvious and few people speak of the terrestrial environment as an entity. Aquatic environments are less familiar to the average person and the differences among them are less obvious, hence the more frequent use of the term *the* aquatic environment, and the tendency to generalize about them and to lump them together as a single entity.

Fresh waters inhabited by vascular plants can be categorized in numerous ways, using different chemical or physical parameters as bases of classification. A commonly used and useful classification is the division into flowing waters and standing waters. As with most other classifications of waters, the dividing line between flowing and standing is somewhat arbitrary. In reality, there are very few natural or man-made bodies of water with no flow at all; and, even in the most stagnant pool, there are currents created by the wind. It is, however, relatively easy to place most bodies of water in one category or the other—flowing or standing. For the most part, the current is clearly a major factor in controlling the conditions of life in a particular location or it is not.

STANDING WATERS

There is a great deal of variation within each of the two major categories. Standing waters, for example, include such diverse bodies of water as lakes, ponds, reservoirs, innundated paddy fields, flooded sand pits, flooded quarries, swamps, marshes, bogs, canals and ditches with negligible currents, sewage lagoons, oxbows, and even large coves and backwaters along slowly flowing rivers. It should be obvious that the conditions for life encountered in each of these types of waters will be unique.

A bog, for example, is entirely different from a swamp. A bog is deep, cold, acidic, and without an inlet or outlet. A swamp is shallow, warmer, may vary in pH, and frequently has a slow, but definite, flow of water through it. Ponds are typically shallow with gently sloping sides and extensive littoral areas, while abandoned, flooded quarries are deep with vertical or nearly vertical sides and no littoral areas at all.

Each of these types of habitat has an association of plants that is adapted to living under the particular conditions that prevail there. Some plants are very adaptable and will be found under a wide variety of conditions; others are very exacting in their requirements and will grow only under narrowly defined, specific conditions.

FLOWING WATERS

Conditions for life in flowing waters also vary dramatically from one water body to another. One is immediately aware of the major differences between large, slow, muddy rivers and small, swift, clear, mountain streams. There are also man-made canals built for drainage, irrigation, and navigation; there are intermittent streams which flow for only part of the year; and there are springs, which are characterized by their consistency in temperature, water chemistry, and flow rate. As with standing waters, each category of flowing water may have its own characteristic flora, but some species of plants are adaptable and will be found in a variety of habitats.

FACTORS AFFECTING AQUATIC HABITATS

Variations also occur within each of the subcategories mentioned above. It is obvious that all lakes are not alike and that all springs are not alike; innumerable variations occur. A few of the more important factors that contribute to these variations are discussed below.

Local Geology and Topography

The nature of the watershed in which a body of water lies cannot be overemphasized. Every aspect of a watershed affects the nature of the aquatic habitats within its boundaries. The size of the watershed affects the amount of run-off, which in turn affects the amount of dissolved and particulate matter entering a lake or stream. The slope of the watershed affects the rate of run-off, which partially determines the degree of erosion, and this also affects the amount of particulate matter entering the water. The nature of the soil and the underlying geologic formations are the primary factors determining the water chemistry, and therefore the productivity, of any body of water. Among other parameters, soils determine the concentrations of many plant nutrients in the system, the hardness of the water, the acidity of the water, and, through erosion, the physical nature of the substrate. Soil type also influences the degree of turbidity of the water, which is extremely important to submerged plants. The effect of soils on water chemistry is exerted not only through run-off, but through water that percolates into the soil, is modified by the soil, and ultimately enters a lake or stream as ground water.

Local Climate

The effect of local climate on aquatic plants is almost as profound as it is on terrestrial plants. The quality and quantity of insolation reaching a plant beneath the surface of the water is controlled by numerous complex factors and will be discussed in detail in Chapter 4. Day length and length of growing season are important considerations, just as they are for terrestrial plants. The angle at which the sun's rays strike the earth are far more important to aquatic plants than to terrestrial ones because it affects the amount of light that will reflect from the water's surface.

Water temperature is also affected by local climate and is a major factor in determining global distribution of aquatic plants. Daily fluctuations in temperature are not as great in aquatic habitats as in the air because of the high specific heat of water. Seasonal fluctuations are also moderated in respect to air temperatures, so that submerged hydrophytes are never exposed to the same extremes of heat and cold as are nearby terrestrial plants. Even floating and emergent forms may be offered some protection from rapid temperature fluctuations by their proximity to the water.

Rainfall is another climatic variable that affects aquatic habitats. The total rainfall during the year is obviously a major factor, but the seasonal distribution of the rain may be just as important or even more so. A habitat that is wet during part of the year and dry during another part of the year requires special adaptations on the part of plants and animals living there.

Human Activities

Almost any human activity within a watershed, if continued long enough, can affect the aquatic habitats within that watershed. Some activities are intentionally designed to affect water. For example, water levels are manipulated for flood control purposes and for gravity-fed irrigation systems. Flow rates are adjusted in streams, and water is diverted for use in other locations. Water is stored behind dams for later use, which affects the downstream flow as well as the reservoir area. Pesticides are added to water intentionally to control weeds or insects. All of these activities obviously have some impact on aquatic habitats.

Other human activities, while undertaken for unrelated purposes, may have enormous impact on water and aquatic habitats. For example, construction, mining, lumbering, and agriculture all disturb the soil and the natural vegetation and tend to increase erosion, which increases the rate at which soil enters and accumulates in lake basins. This contributes to a more rapid filling-in of lakes, increases turbidity, usually increases the levels of certain dissolved plant nutrients, and may change the physical nature of the bottom sediments.

Urbanization, with its accompanying blacktop, concrete, and rooftops, increases the rate of run-off during and after rain. This results in greater water level fluctuation and less ground water seeping into lakes and streams over a long period of time.

Human activities also result in the advertent and inadvertent addition of various chemical substances to aquatic habitats. Domestic sewage, even if well treated beforehand, has an impact if added in sufficient quantity. Septic tank effluents are frequently overlooked as sources of nutrients in lakes that are surrounded by homes. Industrial wastes of many kinds, including acids, dyes, heavy metals, and fibers also find their way into watercourses. Fertilizers and pesticides from farms, home lawns, golf courses, and parks may also enter bodies of water unintentionally, particularly if they are used improperly.

These are but a few of the many ways in which human activities in a watershed may affect aquatic habitats and the plants living in them. It is not necessary that the activity be in or near the water. Activities miles away may begin a chain reaction that ultimately results in a change in a lake, pond, or stream.

DIVERSITY WITHIN A SINGLE BODY OF WATER

Even within a given lake or stream, variations in chemical and physical environments occur. For example, there are differences in flow rate between the bank and the center of a stream and there are differences in flow patterns just upstream from a log or boulder and just downstream from the same log or boulder.

Within a single lake there are differences in light and temperature between top and bottom and between the shoreline and the center of the lake. Sheltered coves may be calm while the open center of the lake, exposed to the wind, may be turbulent. These are a few examples of the subhabitats or microhabitats which may exist within a lake or pond.

Aquatic environments thus constitute an exceedingly diverse group of habitats that should not be thought of as being basically the same with only minor variations. In reality, the only thing they have in common is an abundance of water.

2

Factors Affecting Plant Life in Lakes and Ponds

THE TRANSIENT NATURE OF LAKES

Lakes, ponds, reservoirs, and other standing bodies of fresh water are in a constant state of change. With the exception of those that are artificially cleaned and maintained by man, all lakes and ponds go slowly but continuously through a filling-in process, which eventually leads to their extinction. This is a completely natural and universal occurrence.

Lake basins are filled in by material from various sources. A prime source is sediment that washes in from the watershed as a result of erosion. The nature of the material washing in and the rate of sedimentation vary with the type of soil and vegetation in the watershed, the size and average slope of the watershed, and, to an ever more important degree, human activities in the watershed. The nature and depth of the sediments has a profound effect on the type and amount of vegetation that will exist in a given lake or pond.

Another source of sediments is the shoreline of the lake or pond itself. Shorelines are eroded by wind and wave action and in some cases this can be a substantial source of sediments. Figure 2.1 shows a very small pond whose shoreline has eroded severely and filled in the basin to a point where the pond is useless for most purposes. The severity of shoreline erosion will vary with the type of soil and the type of vegetation on the shore as well as with the amount of wind reaching the water surface. Small ponds in open terrain such as the American Midwest, are frequently protected by planting rows of trees on the up-wind side to serve as windbreaks. By preventing shoreline erosion, these windbreaks slow down the rate of sedimentation and help to keep the water from becoming muddy and turbid.

Remains of dead organisms, both plant and animal, also contribute to the sediments. Dead plant tissue tends to accumulate because there is frequently not enough nitrogen present for the complete decomposition

Fig. 2.1. Shoreline erosion in a small pond. The original shoreline was at the top of the small stone wall and the tile pipe was buried.

of cellulose. As a result, the carbon-rich, nitrogen-poor remains accumulate year after year. The source of the plants may be the pond or lake itself or it may be the surrounding land area. Dead leaves, for example, blow into the water where they settle to the bottom and contribute to the accumulating debris. The siliceous shells of diatoms also persist and add to the sediment.

The remains of dead animals tend to decompose more rapidly and more completely than plants because they lack cellulose and generally have a higher nitrogen-to-carbon ratio than plant tissues. Some animals do contribute substantially to the filling-in process, however. Calcareous mollusk shells, for example, persist under non-acid conditions and form thick deposits on the bottom of certain bodies of water.

The actual rate of filling in varies tremendously because of differences in the variables discussed above. The life spans of some deep lakes are measured in hundreds of thousands of years while others become completely silted in and become marshes in a matter of a few hundred years. Lake Mead, the lake formed behind Boulder Dam, is predicted to have a life of only 150 years because of the heavy silt load carried by the Colorado River (Stevens 1946). Lake Constance, on the Rhine River, is expected to have a life of approximately 12,000 years, half of which is already gone (Strom 1928). Some shallow farm ponds may be become silted in and useless, as far as their original purpose is concerned, after as little as 20 years.

LAKE CLASSIFICATION

Classification Based on Origin

Lakes may be classified on the basis of their origins, and these origins may affect conditions of life for aquatic vegetation growing in them. Several categories of lakes, based on their origins, are given below.

Tectonic Lakes. These lakes are formed as a result of some kind of deformation in the earth's crust — usually associated with earthquakes and faults rather than a slow bending of the rocks. Examples are Loch Ness in Scotland and Lake Tanganyika in Africa. Both are old faults, which eventually filled with water. Tectonic lakes usually start as long, narrow, deep lakes with rocky basins and rocky shorelines. The material washing in from the surrounding watershed tends to be coarse and infertile. This, combined with the great depth and lack of nutrients, results in lakes whose initial productivities are low, and which are colonized by relatively few aquatic plants until they age and accumulate nutrients and a richer, softer substrate.

Volcanic Lakes. Volcanic lakes are formed when craters of old volcanos fill with water. In many respects, conditions for life in these lakes are much like those in tectonic lakes. Substrates are hard and coarse and inhospitable to most rooted plants, and the lakes are deep with steeply sloping sides. With time, finer sediments and plant nutrients accumulate, and these lakes become more productive and support larger standing crops of aquatic plants.

Vulcanism may also result in the formation of new lakes when a lava flow hardens and dams an existing depression in the earth's surface. Initial conditions in such a lake will vary with the nature of the depression at the time it was flooded.

Glacial Lakes. Glaciers can also form lakes and ponds in two distinctly different ways. One way is for the glacier to gouge out a deep basin as it advances, and then to leave a natural dam across the mouth of the basin when it retreats. This dam is called a moraine and consists of material the glacier pushed in front of it while it was advancing and material that was incorporated into the glacier and released when the ice melted. Lakes formed in this manner are likely to have many of the biological characteristics of tectonic lakes and volcanic lakes.

Glacial lakes can also be formed when large blocks of ice detach from a glacier and fall by the wayside. When the main glacier begins to melt, rocks, gravel, and debris that had been incorporated into its ice, wash out to the sides and become somewhat separated by size. The largest rocks remain close to the melting glacier but the finer materials are carried greater distances by the melt water. These finer materials such as silt and clay can fill in the spaces around the detached blocks of ice or even bury them. When the blocks finally melt, they leave water-filled depressions called kettle holes. Kettle holes are common lake types in some parts of the world. They are relatively shallow and their basins contain more silt and clay than the basins of the types of lakes discussed above. As a result, they often have more vegetation and higher productivity than the other types.

Solution Basins. Numerous small lakes are formed in limestone areas when subterranean water dissolves the limestone, leaving underground caverns whose roofs eventually collapse, forming a depression known as a sink hole. These holes fill with water and become small lakes. Sink holes are generally productive because plants are able to use the dissolved limestone as a source of carbon dioxide for photosynthesis.

Basins Formed by Streams. Lakes may form in abandoned stream channels by several processes. In one familiar process, a loop of an old, meandering stream in a flat river valley is cut off when the stream cuts across in a more direct route as illustrated in Fig. 2.2. The abandoned loop gets blocked off at both ends by the deposition of sediments and becomes a type of lake known as an oxbow. Because of locations in old, fertile, river valleys, oxbows are among the most productive natural lakes in the world. Lakes can also form in stream channels in other ways. For example, a temporary channel may form during a period of heavy flooding and when the flood subsides the new channel will remain filled with water.

Man-Made Lakes and Ponds. Artificial bodies of water built by man vary from farm ponds less than, 0.2 hectares (0.5 acres) in surface area to enormous lakes such as Lake Kariba on the Zambezi River in Africa. This lake is 281 km (175 mi) long, 80 km (50 mi) wide, has a surface area of 445,000 ha (1,100,000 acres), an average depth of 45 m (150 ft), and a maximum depth of 114 m (375 ft).

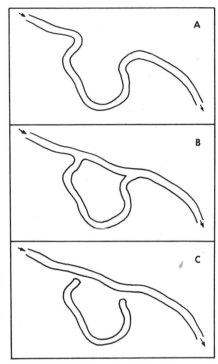

Fig. 2.2. Steps in the formation of an oxbow lake. (A) An old meandering stream with a large loop. (B) A new channel is cut across the neck during a flood. (C) The loop is cut off from the new channel by the deposition of sediment and becomes a semi-circular oxbow lake.

Most man-made lakes are created by damming a stream at a strategic point so that the water backed up by the dam will be contained in a natural valley or basin. Small ponds may ocassionally be created where no stream exists by digging a basin that intercepts the water table. Such a pond is, in a sense, nothing but a wide, shallow well.

Lakes created by the flooding of river valleys are usually shallow, compared to tectonic or volcanic lakes, and they are often extremely productive for a few years following their formation owing to organic and inorganic nutrients in the top soil and the vegetation that was innundated. Furthermore, the material washing into such lakes consists of rich silts and clays. Agriculture is often practiced in these watersheds so that productivity increases rapidly with time.

Classification Based on Productivity

During the past 15 years, the word *eutrophication* has become familiar to every reader of newspapers and popular magazines in the United States. One could not read about the controversy over phosphates in detergents and not be exposed to the word. In the mind of the general public, eutrophication is a process, initiated by humans, that results in massive blooms of algae and the suffocation of fish and other aquatic animals.

The words *eutrophic* and *oligotrophic* were first used in 1907 to describe the water in German peat bogs (Hutchinson 1973). Later the terms were used to describe entire lakes, including the water, substrate, and biota. The definitions are now sometimes expanded to include the lake plus its drainage system and includes concepts of nutrient availability and nutrient cycling as well as the mere presence of nutrients. One classification based on productivity is presented below.

Oligotrophic Lakes.   The classical oligotrophic lake is characterized by very low levels of nutrients and clear water resulting from little plankton growth and little silt or clay in suspension. The morphometry of the basin is typically deep with steeply sloping sides and cold water. Primary productivity is low, as is the standing crop of aquatic plants, including algae.

Eutrophic Lakes.   The classical eutrophic lake is exactly opposite from the classical oligotrophic lake in all respects. It is rich in organic and inorganic nutrients, the water is turbid with plankton and/or suspended soil particles, the lake basin is shallow with gently sloping sides, and the primary productivity and standing crop are both high. In extremely eutrophic lakes, algae is often the predominant form of plant life.

Mesotrophic Lakes.   As limnologists applied this classification system to more and more lakes and its use spread to areas outside of Europe, it became apparent that all lakes did not fit neatly into one of the two categories above, and so a third category, mesotrophic lakes, came into usage. Mesotrophic lakes are intermediate between oligotrophic and eutrophic lakes in productivity. The standing crop of vascular plants may be greater in mesotrophic lakes than in eutrophic lakes because there is less competition from algae.

Dystrophic Lakes.   Dystrophic lakes are a special category of lakes whose waters and substrates contain very low levels of carbonates and other mineral nutrients but whose waters are stained brown with humic, tannic, and other organic acids. The pH of these lakes is quite low as so is their productivity. Aquatic plants that do occur in them form a distinctive association of acid tolerant species.

Eutrophication

As lakes mature they become more and more eutrophic. Even deep tectonic lakes that are extremely oligotrophic when they are formed, become eutrophic with the passage of time. This phenomenon is completely natural and is occurring in every lake all the time. Human activities do not create eutrophication, they merely increase the rate at which it occurs. This does not mean that eutrophication accelerated by man is desireable or that it should be ignored; it only means that some degree of eutrophication will occur with or without human interference.

LAKE STRATIFICATION

Lakes in temperate parts of the world undergo an annual cycle of stratification and destratification, which has a profound effect on chemical and physical conditions in the lake. In order to understand this phenomenon, it is necessary to consider some of the unique physical properties of water that set it apart from all other liquids.

1. *Compressibility.* Within the range of pressures encountered in lakes, water is, for all practical purposes, incompressible.

2. *Temperature—Density Relationship.* Most liquids contract and become more dense as they cool. Water has the unique quality of having its maximum density at 4°C (39.2°F). As the temperature rises above *or falls below* that point, water expands and becomes less dense.

3. *Specific Heat.* Water has a higher specific heat than all but a very few substances such as liquid hydrogen, liquid ammonia, and helium. Compared to other substances, a great deal of energy is required to raise the temperature of a unit weight of water 1 degree. By definition, it requires 1 calorie to raise 1 gram of water 1°C.

4. *Latent Heat of Fusion.* The latent heat of fusion of water is 80 times as great as its specific heat. This means that it takes approximately 80 calories to convert 1 gram of ice at 0° C to 1 gram of liquid water with no change in temperature.

5. *Thermal Conductivity.* The thermal conductivity of water is low. If the lower levels of lakes were restricted to thermal conductivity as a source of heat from above, lake environments would be radically different than they are.

The course of events during a typical year in a temperate zone lake is described below.

Spring

Before the ice cover melts in the spring, the temperature at the interface of the ice and water is 0°C (32°F). (Because of the latent heat of fusion, water in both the liquid phase and solid phase can exist at the same temperature.) The water temperature at the bottom of the lake is 4°C

(39.2°F), because water at that temperature is denser than at any other temperature and will sink to the bottom. When the ice melts and the temperature of the surface water begins to rise above 0° C, it becomes more dense and sinks, to be replaced by cooler water from below. This process continues until the entire lake is 4° C from top to bottom. Since water is essentially incompressable, density differences due to pressure are not a factor, and the lake water is now of equal density from top to bottom.

Because there is no density gradient, winds blowing across the surface create currents that are forced all the way to the bottom by the shape of the lake basin (Fig. 2.3 upper diagram). The entire volume of the lake is mixed thoroughly at this time, using the energy of the wind. Water at the surface that is warmed slightly by the sun and warm air is driven down by the wind and mixed with the deeper, cooler layers so that a uniform temperature is maintained from top to bottom. In this way heat is carried down into the lake at a much faster rate than would occur by thermal conductivity. This mixing is frequently referred to as the spring turnover.

Summer

As air temperatures continue to rise and the sun becomes stronger, the upper layers of the lake absorb more and more heat. This heat is driven down to the lower portions of the lake by the action of the wind. Eventually, on a warm day with little wind, the upper layer of water will heat significantly above the layers immediately below. Rising temperatures now result in water of lower density because the temperature is getting farther from 4°C. If several hot, calm days occur in a row, a layer of warm, light water will float upon the cooler, denser water below. Under these conditions the wind-induced currents no longer reach the bottom of the lake because they cannot penetrate the dense water below. Instead, they return along the top of the denser layer that serves as a barrier.

As the summer progresses and the upper section gets warmer and warmer, the density difference becomes greater and greater and even strong winds are no longer able to drive the currents all the way to the bottom and effect thorough mixing. A stable condition exists known as summer stratification. Under these conditions the lake consists of two separate bodies of water: an upper, warm, well-aerated body of water floating on top of a lower, cooler body of water that becomes oxygen deficient as the summer progresses. The upper layer is called the epilimnion and the lower layer is called the hypolimnion. A thin layer between them is called the thermocline and is characterized by temperatures that decline very rapidly with small increments of depth. A diagram of a stratified lake is shown in Fig. 2.3 (summer).

Autumn

Declining air temperatures in the autumn cool the epilimnion so that its temperature and its density get closer and closer to those of the hypolimnion. Eventually they will be similar enough so that a strong wind will penetrate the thermocline and mixing will again occur all the way to the bottom (see Fig. 2.3). Complete mixing soon results in uniform temperatures and uniform densities from top to bottom. This phenomenon is known as the fall overturn.

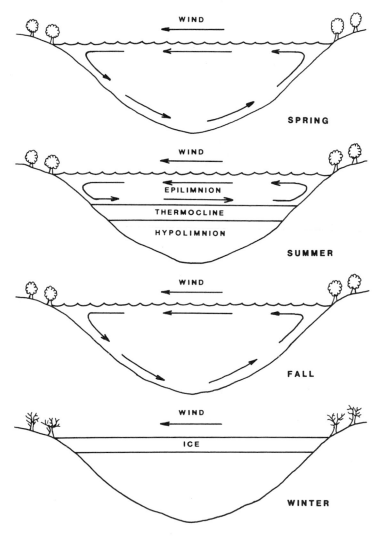

Fig. 2.3.   Seasonal circulation patterns in a temperate region lake.

Winter

As soon as the surface water is cooled below 4°C, it becomes lighter than the warmer water below. The colder it gets, the lighter it gets and the greater is its tendency to stay at the top. Finally, the same phenomenon occurs as that which occurred in the spring or early summer; a temperature stratification with lighter water floating on heavier water. There is a major difference, however. In the winter the light water floating on top is colder than the heavier water at the bottom. This is the reason that lakes freeze from the top down instead of freezing at the bottom first. As soon as the ice cover forms, the lake becomes completely stagnant until the ice melts again in the spring.

This annual cycle has a profound effect on plants that grow in such lakes because, as we will see in later chapters, this cycle affects the availability of oxygen, carbon dioxide, and many dissolved solids.

BIBLIOGRAPHY

HUTCHINSON, G.E.   1978.   Eutrophication. Sci. Am. **61**, 269–279.

STEVENS, J.C.   1946.   Future of Lake Meade and Elephant Butte Reservoir. Trans. Am. Soc. Civil Eng. **111**, 1231. *Cited in*: Sawyer, C. N. 1966. Basic concepts of eutrophication. J. Water Pollution Contr. Fed. **38**, 737-744.

STROM, K.M.   1928.   Production biology of temperate lakes: a symposium based upon recent literature. Int. Rev. Ges. Hydrobiolog. Hydrogr. **19, 329- 348.** *Cited in*: Welch, P. S. 1952. Limnology, 2nd. Ed., McGraw-Hill Book Co., New York.

3

Factors Affecting Plant Life in Flowing Waters

Plants living in flowing fresh waters are subjected to a constant current, but other than that, there is no common factor universal to all flowing-water environments. As much variation can be found among bodies of flowing water as can be found among bodies of standing water.

CURRENTS

Currents in natural bodies of flowing water vary from those in torrential mountain streams to those in large, slow meandering rivers. Even within a single stream a wide variation in current may be observed. As a general rule, the upper portions of a stream, near the headwaters, are on steeper terrain and have more rapid currents than the lower portions of the same stream, nearer to the mouth, which tend to be on flatter terrain and have slower currents.

Local conditions may alter the current radically in small areas. For example, a log or a boulder in a stream will create a turbulence on the upstream side and an eddy on the downstream side, which make conditions for life in those small areas quite different from conditions just a few feet away. Topographic features such as cliffs also modify the course and speed of the current and affect the degree of turbulence. An extreme example would be a waterfall. The area just above the falls, the face of the falls itself, and the pool at the base of the falls, all represent distinctive habitats that differ from one another.

Even within a straight stretch of stream with no bends or obstacles, current velocity is not completely uniform. The current is slowest near the bottom and near the banks due to friction between the water and the stream bed. Surface tension retards the flow somewhat at the surface. As a result, current flow varies at different points in a cross-sectional view of a stream as represented in Fig. 3.1. It should be understood that this diagram represents a highly theoretical, ideal situation. The flow in a

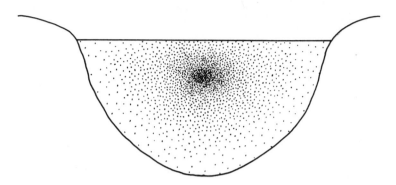

Fig. 3.1.   Diagrammatic cross-section of a stream, showing speed of current under hypothetical, perfectly symmetrical conditions. The greater the density of the shading, the more rapid the current.

natural stream would never come close to this degree of symmetry because of irregularities in the stream bed, winds, and other factors.

STRATIFICATION

Flowing waters do not stratify as standing waters do. Even the slowest currents are sufficient to mix the water from top to bottom as it flows downstream. This mixing also results in relatively uniform chemical characteristics from top to bottom at any given location on the stream.

CHANGES FROM SOURCE TO MOUTH

Numerous physical, chemical, and biological changes will be observed in a stream as it is followed from its source to its mouth. The amount of water being carried by the stream increases as a result of inflow from tributaries, ground water, and runoff. At the same time the velocity of the current diminishes because the terrain tends to become flatter farther from the headwaters. In order to carry the increased volume of water with a slower current, the cross-sectional area of the stream must increase. This is usually manifested as an increase in both depth and width of the channel.

Chemical changes in the water occur as the stream passes through different soil types and different geologic formations. In addition, tributaries may drain areas of different soil and rock types than those found in the area drained by the main channel. Where these tributaries empty into the main stream, abrupt changes in water chemistry may occur.

CHANGES WITH TIME

Rivers and streams, like lakes, evolve with the passage of time. The swift-flowing upper reaches become shorter as the slow-flowing lower reaches encroach further upstream and become longer. As the river basin wears down and becomes flatter the stream meanders more and evolves from a relatively straight course to a winding course with many bends and loops. In the late stages of evolution some of these loops may get cut off when the river is in flood stage and the isolated loop becomes an oxbow lake (see Chapter 2).

As time passes, the stream also cuts deeper and deeper into the earth's crust. This occurs because of the constant scouring action of the current. It represents the exact opposite of the process that occurs in lakes and ponds. In non-flowing waters, sediments accumulate and the depth gradually decreases. In flowing waters, not only are sediments washed away, but the stream bed itself is eroded and the depth increases. The stream may expose new geologic formations as it cuts deeper into the earth, and this may result in major changes in its chemical composition. If, for example, it cuts into a limestone deposit, the pH and the dissolved calcium levels would both rise.

The flood plains along the lower stretches of the river become enriched with silt and organic matter, which wash downstream and become deposited during times of flood when the stream overflows its banks. Other silt and organic debris are carried all the way to the mouth of the stream and drop out of suspension at the point where the stream enters the ocean, a bay, or other standing water body. This sediment accumulates and under the proper conditions, forms a low, fertile, land area known as a delta.

The rate at which all of these changes occur is very variable and depends on numerous factors such as the nature of the soil and underlying materials, the topography of the watershed, and the amount and distribution of rainfall.

CLASSIFICATION OF STREAMS

The classification of streams has not reached the same degree of sophistication and precision as has the classification of lakes. This is due both to the fact that streams have not been studied as intensively as lakes and to the inherent nature of the streams themselves. Since a stream may vary so greatly between its headwaters and its mouth, no single classification will categorize all of the conditions found along its course. Streams are, therefore, classified in only very general terms such as "young" or "old." In order to convey more precise information, a written description is required rather than a mere categorical name.

PLANTS OF FLOWING WATERS

For the most part, rivers and streams have the same species of plants growing in them as those found in standing waters in the same geographic area. With one exception, which will be discussed below, there is not a separate, distinct group of plants that can be identified as being "river plants" and which would be out of place in standing waters. Weakly rooted species and free-floating species are, of course, restricted to slower moving waters where they are not subject to being swept away by the current. Even in waters of moderate current, however, plants such as duckweeds and watervelvet may be found growing among debris, emergent plants, and overhanging branches along the stream margins. Water hyacinth is a common pest in flowing waters even though it is free-floating and not anchored to the substrate.

Podostemaceae

The family Podostemaceae (including Tristichaceae of some authors) is the exception to the statements made in the paragraph above. These plants are restricted to flowing waters, and for the most part they are restricted to rapids, waterfalls, and other areas of rapidly flowing, turbulent water. There are 46 genera and approximately 260 species widely distributed throughout the tropics and with one genus extending into the subtropics of North America. Several of the species that have been described are known only from a single cataract or a single waterfall and apparently do not exist anywhere else in the world.

Plants of this family are superbly adapted for life in rapidly flowing water. There are no recognizable stems or roots. The plant body consists of a thallus, which is probably composed of both stem and root tissue. It is flattened in form and excretes a cement that fastens it to the bare rock surfaces on which the plants grow. The leaves are very flexible and offer little resistence to the current. There is no internal aeration system (lacunae), as is typical of virtually all other submerged plants. Such a system is apparently unnecessary because the turbulent waters in which the plants live are highly aerated and oxygen and other dissolved gases are transported to all parts of the plant in sufficient quantity by diffusion alone.

Most members of the family are red in color and many resemble marine algae, mosses, or lichens in general shape and habit of growth. None have any economic importance, but some play important roles in the life cycles of aquatic insects. They have been studied in some detail as biological curiosities.

4

Light in Underwater Environments

Of all major environmental parameters, the two that differ most between terrestrial and submerged environments are water and light. The differences in water availabilty are obvious. The differences in light quality and intensity between the two types of environments are not nearly as obvious but are of tremendous importance.

Light affects many biological processes and activities. The time of spawning of some fish is determined by photoperiod; feeding activities of some fish and vertical migration of some zooplankters are controlled by light intensity; and prolonged exposure to certain parts of the spectrum can be fatal to some microorganisms and fish eggs.

Our primary interest in subaquatic light is through its role in photosynthesis. Every green plant is dependent on light as an energy source for photosynthesis, and all freshwater food chains have green plants as their bases. For this reason alone, light is of enormous importance in the underwater environment, and deserving of detailed study.

FATE OF INCIDENT RADIATION

Before discussing underwater light we should review some basic physics and find out what may occur to light striking the surface of a lake, pond, or stream. Figure 4.1 shows in a simplified way the path a ray of light (the incident ray) striking the water surface can follow. Upon striking the surface at point A, a portion of the ray is reflected back into the air; the remainder penetrates the surface and is refracted by the difference in the optical densities of the air and water.

The reflected ray is lost to the underwater environment and will not be considered further (although it could be reflected back to the surface again by suspended particles in the air). The penetrating ray passes

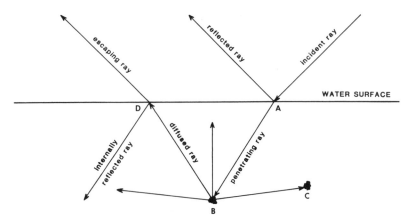

Fig. 4.1.   Possible paths of a ray of light (the incident ray), striking a water surface.

through the water, and parts of it are immediately absorbed by the water itself and by dissolved substances in the water. That portion of the penetrating ray not absorbed may eventually strike a solid particle suspended in the water (B). Part of the energy will be absorbed by the particle and part will be diffused and reflected back into the water as numerous diffused rays. The diffused rays will also be absorbed by the water and dissolved substances until they strike a solid (C) or the surface (D), where a portion will escape back into the atmosphere and a portion will be reflected downward.

Ultimately, all of the energy in the original incident ray will either escape back into the atmosphere or will be absorbed by the water, by dissolved substances in the water, or by solids in the underwater environment. These solids may be minute suspended particles, both animate and inanimate, larger plants and animals, or the substrate of the lake basin or stream channel.

Surface Loss

Surface loss has two distinct components: light that is reflected from the surface on initial impact and which never gets into the water, and light that penetrates the surface but is scattered back into the atmosphere after striking solid materials under the water. The percentage of incident light lost on initial impact varies with the elevation of the sun, the condition of the sky (clear or cloudy), and the condition of the water surface (rough or smooth).

Total surface loss (initial loss plus escaping underwater light) on a cloudy day where all of the incoming light is diffuse averages about 6% of incident radiation (Davis 1941). On sunny days with a calm surface, total surface loss varies from approximately 3–17% depending on the angle of the sun. Figure 4.2 shows that surface loss is not equal for all portions of the spectrum. The solid line represents surface loss for total light and the dotted line represents surface loss for the red portion of the spectrum only. It can be seen that surface loss for red light is less than that for total light at zenith sun angles up to 76°. At greater angles, when the sun is closer to the horizon, the percentage loss of red light is greater than that of total light. Thus, the angle of the sun not only affects the quantity of light found below the surface, but the spectral composition of the light as well.

Quantitative Light Transmission By Water

Pure Water. Light absorbtion by pure water varies markedly with the wavelength of the light. Figure 4.3 shows the percentage of light of different wavelengths absorbed by passing it through 1 m (3.28 ft) of especially distilled water of very high purity. Maximum absorption occurs at 7600 Å (extreme red). Passage of light of this wavelength through 1 m of pure water results in the absorption of 91.4% (transmission of only 8.6%). Minimum absorption occurs at 4730 Å. Passage of light of this wavelength through 1 m of pure water results in the absorption of only 0.49% (transmission of 99.51%).

The amount of natural sunlight absorbed by passage through 1 m of pure water varies somewhat because the spectral composition of natural sunlight varies with atmospheric conditions. Water vapor and dust in the atmosphere, for example, function as natural filters, selectively reducing certain wavelengths in relation to others. Under all atmospheric conditions, however, less than 50% of incident natural sunlight penetrates 1 m

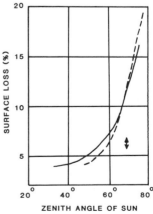

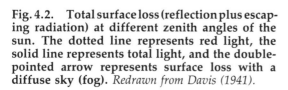

Fig. 4.2. Total surface loss (reflection plus escaping radiation) at different zenith angles of the sun. The dotted line represents red light, the solid line represents total light, and the double-pointed arrow represents surface loss with a diffuse sky (fog). *Redrawn from Davis (1941).*

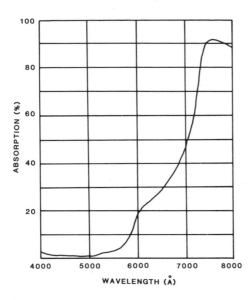

Fig. 4.3.   Absorption of light of various wavelengths by 1 m of distilled water. *Redrawn from James and Birge (1938).*

of pure water. It is obvious, therefore, that submerged aquatic plants, unless they are growing in very shallow water, receive far less light than plants growing in unshaded terrestrial environments. Even the clearest of waters has a powerful shading effect.

Natural Waters.   Light transmission in natural waters is affected not only by the absorption of the water itself, but by dissolved materials (stains) and suspended solids (turbidity). Even clear lakes and streams have some dissolved and suspended solids, which makes them poorer transmitters of light than they would otherwise be.

The greatest absorption of light in natural waters (on a percentage basis), occurs in the upper layers. Deeper layers transmit the available light with ever greater efficiency for two reasons. First, there is more plankton in the upper layers to absorb radiation. Second, those wavelengths that are most easily absorbed are selectively removed by the upper layers. As a result, the light that penetrates to the deeper layers is composed to an ever greater degree of those wavelengths that are transmitted best.

Birge and Juday (1929, 1930, 1931, 1932) measured light transmission in more than 500 lakes in Wisconsin. Transmission curves for three selected lakes are presented in Fig. 4.4. All three curves have been corrected for a zenith sun so that they are directly comparable. The small upper curve represents a very turbid lake. At a depth of less than 2 m, 99% of the incident light was absorbed. The middle curve represents a lake of me-

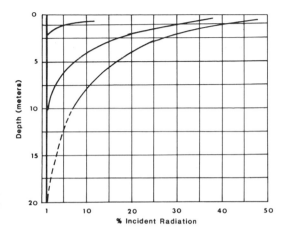

Fig. 4.4. Light transmission curves for three Wisconsin lakes. The dotted portion of the lowest curve represents an extrapolation beyond the actual depth of the lake. *Redrawn from Birge and Juday (1929).*

dium turbidity and color. In this lake, 99% of the incident light was absorbed by a column of water 10 m deep. The lower curve represents the clearest lake that Birge and Juday investigated. In this lake, 99% of the incident light was not removed until a depth of 20 m was reached.

It can readily be seen from this graph that with the sun directly overhead, a plant leaf 1 m below the surface of a lake may receive as much as 45% of incident light or as little as 5%, depending on the clarity of the water. This difference is obviously great enough to be a major factor in controlling which species of plants can survive in a given body of water.

Snow And Ice. During the winter months when lakes and ponds in temperate and northerly climates have covers of ice, and sometimes snow, on them, the amount of light that penetrates into the water is markedly reduced. The actual percentage of transmission is highly variable and is affected not only by the thickness of the ice or snow, but by its condition. Ice may be clear or cloudy and snow may be dry and light, wet and compact, crusted on top, or even so wet as to be transluscent. A few figures taken from a study by Greenbank (1945) will show the impact that ice or snow condition can have on light transmission. In one instance, 19 cm (7.5 in.) of clear natural ice on a lake transmitted 84% of the incident radiation (including surface loss). In another instance, the same thickness of, ice described as being "fairly clear," transmitted only 53% of the incident light. The thickest ice reported on in this study was 61 cm (24 in.) thick and was described as being "partly cloudy." Only 7.6% of the incident light penetrated this cover.

Snow is even more variable in transparency than ice. Light transmission was measured through undisturbed, natural snow by placing the detector of a light meter on a frozen lake and allowing natural snowfall to cover it. Light transmission was measured remotely via electrical leads coming up through the snow. On four seperate ocassions, the percentages of light penetrating 2.54 cm (1 in.) of snow were 29%, 17%, 13%, and 10%. This represents a threefold difference due to the condition of the snow. The lowest percentage of transmission reported was 0.7% which was measured under 25 cm (10 in.) of dry snow.

Qualitative Light Transmission by Water

As we have already seen, the absorption of light by water is not equal at all wavelengths. As a result, the quality of light (its spectral composition), changes as it penetrates into a lake, pond, or stream. Changes result from selective absorption by the water itelf, by stains (dissolved substances) in the water, and by solid particles suspended in the water. Water itself absorbs most efficiently in the red portion of the spectrum, while most natural stains and suspended particulate matter absorb most efficiently in the blue portion of the spectrum.

As a very general rule of thumb, natural waters of high transparency transmit blue and green light best, natural waters of medium transparency transmit yellow light best, and natural waters of low transparency transmit red and orange light best. This is readily apparent from Tables 4.1 through 4.4. These tables present data from four lakes selected from among more than 500 studied by Birge and Juday (1931). For each lake the total light at various depths is broken down and presented as the percentage in each of the following parts of the spectrum: red, orange, yellow, green, blue, and violet. In natural lakes, at depths of 1 m (3.28 ft) or more, the spectrum extends little if at all beyond the visible range, and therefore the figures in the four tables represent total radiation.

Crystal lake (Table 4.1) is an exceedingly clear lake. At a depth of 9 m (29.5 ft) the yellow, green, and blue portions of the spectrum account for 78% of the total, while red is completely absent and orange accounts for only 8%. Black Oak Lake (Table 4.2) is a lightly colored lake. At a depth of 9 m (29.5 ft) in this lake, the red portion of the spectrum is also absent, but so is the violet. Blue is reduced considerably instead of increasing as in Crystal Lake. The yellow and green portions of the spectrum show the greatest percentage increases. Little Papoose Lake is a lake of moderate color and transparency. At a depth of only 3 m (9.8 ft), the violet portion of the spectrum is completely absorbed and the blue accounts for a mere 4%. The red and orange portions of the spectrum, however, account for 61% of the total radiation. Lake Mary (Table 4.4) is the least transparent of all the lakes studied by Birge and Juday, being heavily colored with organic stains. At the very shallow depth of 2 m (6.6 ft) the green, blue, and violet

TABLE 4.1. Spectral Distribution of Light and Percentage of Incident Radiation
Remaining at Various Depths in Crystal Lake, Wisconsin

Depth (m)	Color[a] of Light (% of Total)						Incident Radiation Remaining (%)
	V	B	G	Y	O	R	
1	18	16	15	18	18	15	38
3	16	22	20	22	12	8	24
5	16	20	27	23	10	4	18
7	14	25	27	23	10	1	14
9	14	24	30	24	8	0	10

From Birge and Juday, (1931).
[a] V = violet, B = blue, G = green, Y = yellow, O = orange, R = red.

TABLE 4.2. Spectral Distribution of Light and Percentage of Incident Radiation
Remaining at Various Depths in Black Oak Lake, Wisconsin

Depth (m)	Color[a] of Light (% of Total)						Incident Radiation Remaining (%)
	V	B	G	Y	O	R	
1	11	19	16	20	18	16	34
3	9	14	21	29	18	9	17
5	5	15	24	33	17	6	9
7	2	16	25	37	17	3	5
9	0	5	25	34	16	0	2.5

From Birge and Juday (1931).
[a] V = violet, B = blue, G = green, Y = yellow, O = orange, R = red.

TABLE 4.3. Spectral Distribution of Light and Percentage of Incident Radiation
Remaining at Various Depths in Little Papoose Lake, Wisconsin

Depth (m)	Color[a] of Light (% of Total)						Incident Radiation Remaining (%)
	V	B	G	Y	O	R	
1	3	9	11	29	20	28	17
2	0	7	17	22	26	28	[b]
3	0	4	20	25	28	35	3.1

From Birge and Juday (1931).
[a] V = violet, B = blue, G = green, Y = yellow, O = orange, R = red.
[b] = value not reported.

TABLE 4.4. Spectral Distribution of Light and Percentage of Incident Radiation
Remaining at Various Depths in Lake Mary, Wisconsin

Depth (m)	Color[a] of Light (% of Total)						Incident Radiation Remaining (%)
	V	B	G	Y	O	R	
1	0	3	3	12	28	54	4
1.5	0	0	2	8	35	55	[b]
2	0	0	0	5	40	55	0.5

From Birge and Juday (1931).
[a] V = violet, B = blue, G = green, Y = yellow, O = orange, R = red.
[b] = value not reported.

portions of the spectrum are already completely absorbed. The yellow portion of the spectrum accounts for only 5% of the total, and red and orange together make up 95% of all light.

When interpreting these tables it must be remembered that all portions of the spectrum decrease in absolute intensity with increasing depth. It is only the relative intensity of some portions that increase in relation to others.

LIGHT AND SUBMERGED PLANTS

From the previous discussion it is apparent that submerged plants are subjected to light of lower intensity and different spectral composition than that which reaches the surface of the earth. The deeper a plant is in the water, the lower the intensity will be and the greater the spectral divergence from "normal" will be.

Light Intensity

Most submerged vascular plants are adapted to, and indeed require, low light intensities. Studies on the effects of light intensity on eight species of algae and five species of submerged vascular plants were conducted by Boyd (1975). The results of this work are summarized in Fig. 4.5. The vertical arrow in this figure represents increasing light intensity. The statements on the left of the arrow are summary statements for the eight species of algae tested and the statements to the right of the arrow are summary statements for the five species of vascular plants tested.

It can be seen that maximum rates of photosynthesis for the algae occurred between 20,000 and 35,000 lux, but maximum rates for the vascular plants occurred at much lower intensities (10,000–20,000 lux). Intensities above 20,000 lux usually inhibited photosynthesis in the vascular species. These plants exhibited 50% of their maximum photosynthetic rates at light intensities of only 2000–5000 lux, which is less than 15% of full summer sun. The ability to photosynthesize efficiently under conditions of low light intensity is an obvious advantage to submerged plants.

All submersed aquatic species are not equally adapted to low light conditions, however. In one laboratory study, Blackburn et al. (1961) found that Elodea densa grew optimally at an intensity of 107 lux (approximately 0.3% of full summer sun), and intensities of 1345 lux and higher were highly injurious. Heteranthera dubia, on the other hand, grew optimally at 6350 lux (approximately 18% of full summer sun), or greater.

If the light intensity requirements of more submerged species were known, it is likely we would find a continuum of shade tolerance from

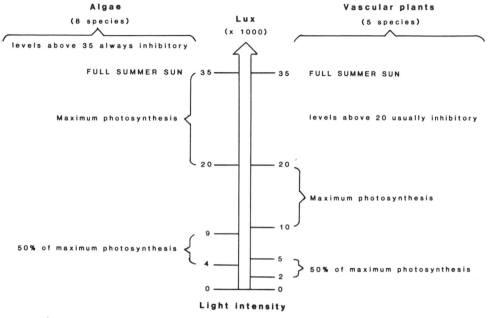

Fig. 4.5. Summary of the responses of eight species of algae and five species of vascular plants to varying levels of light intensity. *Data from Boyd (1975).*

extremely tolerant to moderately tolerent. Most species, however, would be considerably more tolerant to low light than are average terrestrial plants. Under actual field conditions almost all submerged plants are limited to depths of 8 m (26 ft) or less. The greatest depth at which a vascular plant has ever been reported to be growing (as opposed to plants which merely drifted down after having been growing at shallower levels), is 11.5 m (38 ft). This was a specimen of *Potamogeton strictus* found in Lake Titicaca, on the Peruvian-Bolivian border (Hutchinson 1975). Certain mosses and macroscopic algae of the family Characeae have been found living at depths in excess of 100 m (328 ft).

It is generally agreed that light intensity is the limiting factor in determining the maximum depth at which vascular plants will occur in a given body of water. The depth at which respiration exactly equals photosynthesis for a particular plant is known as the compensation depth. No plant can persist for other than short periods of time below its compensation depth because under those conditions it uses stored food materials faster than it produces them, and it eventually starves to death.

It is possible that in an exceedingly transparent lake of great depth, pressure could limit the maximum depth of colonization by vascular plants. The effects of high pressure on aquatic plant growth have been studied only sparingly, but it is known that abnormal pressure can produce abnormal growth. In most lakes, however, light becomes a limiting factor long before pressure does.

Light Quality

Relatively little is known about the effects of varying spectral composition on photosynthesis in aquatic species. Chlorophyll absorbs most efficiently in the violet-blue portion of the spectrum and the orange-red portion of the spectrum. It absorbs least efficiently in the green-yellow portion. The deeper zones of transparent lakes, therefore, may not always have those wavelengths of light that are most efficient for photosynthesis, even though their total light intensity is greater than that at the same depth in less transparent lakes.

Some aquatic plants, however, seem to be adapted to the very wavelengths that predominate at the bottom of deep, clear lakes. Under controlled laboratory conditions *Heteranthera dubia* grew three times as rapidly under green-yellow light (480−630 nm) as it did under daylight lamps of the same intensity. *Elodea densa*, on the other hand, turned yellow and failed under the green-yellow light but thrived under the daylight lamps (Blackburn *et al.* 1961). In each case the plants were grown under light intensities that were near optimum for the species involved. The physiological basis of this type of adaptation is unknown.

BIBLIOGRAPHY

BIRGE, E.A., and JUDAY, C. 1929. Transmission of solar radiation by the waters of inland lakes. Trans. Wisc. Acad. Sci., Arts, Letters. **24**, 509−580.

BIRGE, E.A., and JUDAY, C. 1930. A second report on solar radiation and inland lakes. Trans. Wisc. Acad. Sci., Arts, Letters. **25**, 285−335.

BIRGE, E.A., and JUDAY, C. 1931. A third report on solar radiation and inland lakes. Trans. Wisc. Acad. Sci., Arts, Letters. **26**, 383−425.

BIRGE, E.A., and JUDAY, C. 1932. Solar radiation and inland lakes, 4th Report. Observations of 1931. Trans. Wisc. Acad. Sci., Arts, Letters. **27**, 523−562.

BLACKBURN, R.D., LAWRENCE, J.M., and DAVIS, D.E. 1961. Effects of light intensity and quality on the growth of *Elodea densa* and *Heteranthera dubia*. Weeds 9, 251−257.

BOYD, C.E. 1975. Competition for Light by Aquatic Plants. Agric. Exp. Sta. Circ. 215. Auburn Univ., Auburn, Alabama.

DAVIS, F.J. 1941. Surface loss of solar and sky radiation by inland lakes. Trans. Wisc. Acad. Sci., Arts, Letters 33, 83−93.

GREENBANK, J. 1945. Limnological conditions in ice- covered lakes, especially as related to winter-kill of fishes. Ecol. Monogr. **15**, 343−392.

HUTCHINSON, G.E. 1975. A Treatise on Limnology, Vol. III, Limnological Botany. John Wiley & Sons, New York.

JAMES, H.R., and BIRGE, E.A. 1938. A laboratory study of the absorption of light by lake waters. Trans. Wisc. Acad. Sci., Arts, Letters. **31**, 1−154.

5

Plant Nutrients in Aquatic Environments

Unlike terrestrial plants, submerged aquatic species are continually bathed in a solution of dissolved nutrients. This, coupled with the fact that they are not in direct contact with the atmosphere, makes their nutritional problems somewhat different from those of terrestrial plants. This chapter will focus on the more important of these differences.

THE CONCEPT OF ESSENTIAL ELEMENTS

The term "essential element" has a very precise meaning in relation to plant nutrition. In order to qualify as an essential element a nutrient must satisfy one of the following criteria: (1) in the absence of the element, the plant cannot complete its life cycle, or (2) the element is part of a compound that is absolutely necessary for plant life.

Some elements can be utilized by plants, but they are not essential elements because they can be replaced by others. Sodium, for example, can be utilized by plants in certain biochemical reactions, but it can be replaced by potassium in all cases. Sodium, therefore, is not an essential element. Potassium, on the other hand, is necessary for some biochemical reactions in which it cannot be replaced by sodium (or any other element), and potassium, therefore, is an essential element because it satisfies the first criterion. The terms "functional nutrient" and "metabolism nutrient" have been proposed to describe any mineral element that functions in plant metabolism, whether or not it is an essential element (Nicholas 1961).

Water is necessary for all plant life, and since the water molecule contains hydrogen and oxygen, they are essential elements because they satisfy the second criterion. Similarly, magnesium is contained in the chlorophyll molecule and thus automatically qualifies as an essential element.

Sixteen essential elements are known, and although little work has been done to test the essentiality of nutrients for aquatic plants specifically, there is no reason to believe that the list is any different from that established for vascular plants in general. The 16 essential elements have been divided into two groups, micronutrients and macronutrients, based on the relative amounts required by plants.

The macronutrients are carbon, hydrogen, oxygen, nitrogen, phosphorus, sulfur, potassium, calcium, and magnesium. Even though they are all required in relatively large amounts when compared to the micronutrients, there is a great deal of variation among them in regard to how much plants need. Between 95 and 99.55% of the weight of fresh plant tissue is made up of carbon, hydrogen, and oxygen (Jones 1979). Magnesium, in contrast, makes up only about 0.18% of the weight of a corn plant (Epstein 1972).

The micronutrients are required in smaller quantities. The amounts required are so small that they are frequently called trace elements in the agricultural literature because only traces are necessary for successful crop production. The seven micronutrients are iron, manganese, boron, zinc, copper, molybdenum, and chlorine. Micronutrients are no less essential to plants than macronutrients; they differ only in the amounts required.

CARBON, OXYGEN, AND HYDROGEN

Carbon, oxygen, and hydrogen are unique among the essential elements in that they are obtained from water and the atmosphere. Together, they constitute a very high percentage of the fresh weight of plant tissue because they are the sole elements of water and carbohydrates and they are abundant elements in proteins and plant lipids.

Carbon

Carbon is the basic atom about which all organic molecules and therefore life, as we know it, is built. Terrestrial plants and emergent aquatic plants obtain carbon from the atmosphere in the form of free CO_2, which enters the stomata on the leaves. Plants such as waterlilies, which have floating leaves and large submerged and subterranean organs, have special adaptations for transporting atmospheric gases to these organs (see Chapter 9).

Submersed plants, which are not in direct contact with the atmosphere, must obtain their CO_2 from the supply that is dissolved in the water. This is one of the major differences between the nutritional aspects of terrestrial plants and those of aquatic plants. When CO_2 dissolves in water a certain percentage of the molecules react with the water to form a very complex equilibrium. A simplified version of this equilibrium is shown in Eq. 5.1.

$$CO_2 + H_2O \rightleftharpoons H_2CO_3 \rightleftharpoons H^+ + HCO_3^- \rightleftharpoons H^+ + CO_3^{2-}$$
$$+$$
$$H_2O$$
$$\updownarrow$$
$$H_2CO_3 + OH^- \tag{5.1}$$

It can be seen that carbonic acid, carbonate ions and bicarbonate ions can all be converted to free CO_2 by shifting the equilibrium to the left. Much of the CO_2 in aquatic environments is stored in the form of carbonates and bicarbonates. Limestone in a lake basin or stream channel (or anywhere in the watershed) is thus a potential source of CO_2 for photosynthesis by submersed aquatic plants. For this reason, hard waters are frequently more productive than soft waters.

When plants use CO_2, the equilibrium depicted in Eq. 5.1 will shift to the left, in effect replacing the CO_2 that was used. This is a "passive" reaction on the part of the plant whose only role was the initial removal of one factor on the left side of the equilibrium. There is evidence that many submersed plants have the capability of actively assimilating and decomposing bicarbonates to form carbonates, water, and carbon dioxide as shown in Eq. 5.2.

$$Ca(HCO_3)_2 \rightleftharpoons CaCO_3 + H_2O + CO_2 \tag{5.2}$$

An excellent review of the research that has been done on this phenomenon is given by Hutchinson (1975). The actual mechanism by which the plants utilize the bicarbonate ion is unknown, but the capability is widespread among submersed angiosperms. Species having this ability have a great competitive advantage over those that do not because free CO_2 is frequently a limiting factor to growth in dense stands of aquatic vegetation. The carbonate formed on decomposition of the bicarbonate frequently precipitates on the surfaces of the stems and leaves, giving the plants a coarse, limey texture.

One of the most unique aspects of carbon in aquatic environments is the large and rapid effect that plant metabolism may have on the free CO_2 supply. The utilization of CO_2 in photosynthesis may be so rapid that the supply is reduced to essentially zero under conditions of adequate light.

Carbon dioxide is produced by plant respiration, which results in elevated concentrations in the water during periods of darkness. The carbon supply of submerged plants, therefore, must be considered in terms of the diurnal/nocturnal cycle as well as the total supply over the period of the growing season. The ecological effects of these fluctuations are discussed in Chapter 8.

Oxygen

Land plants, emergent aquatic species, and floating-leaved aquatic species obtain oxygen from the air in the same manner in which they obtain carbon dioxide. Submerged aquatic plants, however, absorb dissolved oxygen from the water in which they live. Dissolved oxygen levels vary from saturation or near saturation in rapidly flowing streams to anaerobic conditions in some highly polluted waters and in the hypolimnion of stratified lakes. Most submerged plants have internal spaces where some photosynthetic oxygen may be stored but the gas in the spaces equilibrates rather quickly with the surrounding water and these spaces are not major source of oxygen for the plant.

Since oxygen does not become chemically bound in the form of inorganic salts in the same way that CO_2 does, it cannot be "stored" in this form nor can it be added to the water from dissolving rocks. Dissolution from the atmosphere is the major source of oxygen in any body of water, and since the degree of mixing determines how much water actually comes into contact with the air, stratification profoundly affects the oxygen economy of a lake.

Photosynthesis in submerged plant tissues is another source of oxygen in underwater habitats. It must be remembered, however, that the very tissues that supply oxygen during daylight hours, remove oxygen from the water during the night or on dark days when light intensity falls below the compensation level and respiration exceeds photosynthesis. Thus, oxygen concentrations, like CO_2, may undergo a 24 hour cycle of increase and decrease.

Hydrogen

Plants obtain all of their hydrogen in the form of water. Although most of the water that enters a plant remains in liquid form, some water molecules are split into oxygen and hydrogen, using energy from the sun. The hydrogen is trapped and used in photosynthesis and the oxygen is released to the surrounding environment. Since aquatic plants, by definition, grow in areas where water is abundant, hydrogen is never a limiting factor to their growth.

PHOSPHORUS AND NITROGEN

Phosphorus and nitrogen will be considered next since these are the two elements most frequently implicated in the acceleration of eutrophication by mankind. Both may be limiting to plant growth in some unpolluted waters and additions of either or both are likely to increase productivity.

Phosphorus

Phosphorus is important in the reproductive growth of plants and in the formation of high energy bonds such as those found in ATP. It is present in natural waters in many different forms, which has caused a great deal of confusion in the interpretation of published data. It is present in dissolved form, adsorbed onto particulate matter suspended in the water, adsorbed onto organic matter suspended in the water, contained in living and dead organic molecules, adsorbed onto soil particles in the sediments, and is present as precipitated inorganic salts in the sediments. With so many forms to consider, one must be very careful in interpreting published results of water analyses and in comparing the results of different studies. One study may be reporting concentrations of one form of phosphorus while another study may be reporting a different form that is not directly comparable with the first. Even dissolved orthophosphate may be present in more than one ionic form, as shown in Table 5.1.

Phosphorus is present in natural waters in extremely small amounts. Oligotrophic lakes frequently have less than 0.001 mg/liter (1 ppb) in solution. Juday and Birge (1931) made phosphorus determinations on 479 lakes in Wisconsin and found an average of 0.023 mg total P/liter and 0.003 mg soluble P/liter. In a much later study (Norvell and Frink 1975), the average total phosphorus content of 23 lakes in Connecticut was determined to be 0.0228 mg/liter. Rounded off, this is the same value reported more than 40 years earlier for the Wisconsin lakes.

TABLE 5.1. Effects of Water pH on the Ionic Species of Dissolved Orthophosphate

Ionic Species	pH Range over Which Species May Occur	pH at Which Approx. 50% of Ions Will Be of Indicated Species
H_3PO_4	1–4	2.0
$H_2PO_4^-$	2–7	2.5
HPO_4^{2-}	7–12	7.5
PO_4^{3-}	12–14	12.5

How are lakes with such low phosphorus concentrations in their waters capable of supporting plant growth? These small amounts should be depleted very quickly during periods of rapid plant growth, especially by algae, which are capable of taking up available nutrients in very short periods of time. The answer to this question is that there are much larger stores of phosphorus in the bottom sediments than in the water at any given moment, and the phosphorus in the sediments is in equilibrium with that in the water. As plants remove the soluble form from the water, more phosphorus is released from the sediment to restore the equilibrium. In this way, far more phosphorus is ultimately available to the plants than that which is in the water at any one time.

The factors that control the equilibrium between sediment and water are numerous and complex. Phosphorus in the sediment exists as relatively insoluble salts and as ions that are bound on clay minerals, organic matter, and hydroxy gels. The release of phosphorus from the sediments is dependent on many factors, but pH and oxygen concentrations are especially important in natural waters. Lowering the pH results in the release of more phosphorus from the sediment to the water. Raising the pH results in more phosphorus being tied up in the sediment in forms temporarily unavailable to plants.

In the presence of oxygen, iron is present in the oxidized state and forms relatively insoluble salts with phosphorus, that precipitate out of solution. Under anaerobic conditions or when oxygen levels are low, iron is reduced and the corresponding phosphorus salts that form are relatively soluble. Ferrous phosphate, in other words, is considerably more soluble than ferric phosphate. In addition, the charge by which phosphorus is bound to hydroxy gels of iron is also a function of the oxidation state of the iron. If ferrous hydroxide is oxidized to ferric hydroxide, there is an additional positive charge by which phosphorus may be bound. As a result, if all other factors are equal, more phosphorus will be found in solution under conditions of low oxygen tension than under conditions of high oxygen tension.

In stratified lakes this means that the oxygen-rich epilimnion will have less available phosphorus than the oxygen-poor hypolimnion. As the summer progresses, the epilimnion becomes poorer and poorer in phosphate as insoluble precipitates and the dead bodies of plankton and larger plants and animals sink into the hypolimnion carrying their phosphorus with them. This, coupled with release from the sediments, results in high concentrations of phosphorus in the hypolimnion by late summer. When the fall overturn occurs and the lake mixes thoroughly, some of this phosphorus is recycled and made available to plants again.

During the winter months when lakes are ice-covered in northern climates, the deeper waters may again become oxygen-deficient, releasing more phosphorus, which is recycled for plant growth at the time of the spring turnover. Thus, phosphorus concentrations in the upper layers of lakes are typically highest in spring and fall immediately following the turnovers.

In unpolluted waters the primary source of phosphorus is the weathering of phosphatic rock from which it leaches very slowly. As a result the total amount of phosphorus in a lake system (including the sediments) changes very slowly with time.

Nitrogen

Nitrogen is, in many respects, the exact opposite of phosphorus. It is important to plants in vegetative growth rather than reproductive growth. It is also an integral part of all amino acids and, therefore, of proteins.

The great reservoir of nitrogen in the world is free nitrogen gas (N_2) in the atmosphere. Nitrogen in this form cannot be used by plants until it is transformed into NO_3 or NH_4 by nitrogen-fixing bacteria, blue-green algae, or, to a lesser extent, by lightning. Although the diatomic gas obviously dissolves in waters of lakes, ponds, and rivers, it has no direct biological significance except to nitrogen-fixing organisms. In the following paragraphs, the word nitrogen refers to fixed forms of nitrogen only.

Unlike phosphorus salts, most inorganic salts of nitrogen are very soluble in water. Furthermore, they are not readily bound onto the exchange complex of clay minerals, organic matter, or gels in the sediments. As a result, a greater percentage of the inorganic nitrogen in a typical aquatic ecosystem is found in the water and a smaller percentage exists as an exchangeable reservoir in the bottom muds, than is the case with phosphorus. Where sediments contain a great deal of organic matter, the organic nitrogen level may be high, and this serves as another reservoir of the element, although this nitrogen is not in rapid equilibrium with the water and is not immediately available to plants.

In a study of Bantam Lake, Connecticut, Frink (1967) estimated that during the growing season, the water in the entire lake contained approximately 10,700 kg (23,600 lb) of nitrogen while the top 1 cm (0.39 in.) of sediment contained 44,000 kg (97,000 lb). Thus, there was roughly four times as much nitrogen in the top centimeter of sediment as in the water. In contrast, the water contained 460 kg (1014 lb) of phosphorus but the upper centimeter of sediment contained 7400 kg (16,300 lb), which is about 16 times the amount found in the water.

Nitrogen concentrations in natural waters are not only considerably higher than those of phosphorus (5 mg/liter is not uncommon for short periods of time), but they also fluctuate more rapidly and within a much wider range. Because nitrogen is generally more soluble and more mobile, it leaches out of soils more readily and therefore washes into and out of water courses with greater ease. In addition, the bottom muds do not serve as a rapid equilibrating system as they do for phosphorus, and thus do not function to maintain a more uniform concentration in the water.

In stratified lakes, nitrogen in the hypolimnion is present primarily as NH_4 when oxygen levels are low. When oxygen levels are high, the fixed nitrogen is present primarily as NO_3.

Although nitrogen may become limiting to plant growth under certain circumstances, most authorities feel that it is less frequently limiting than phosphorus and carbon dioxide. Since nitrogen is not easily bound to exchange sites and since there are few insoluble salts that will precipitate out of solution, the removal of this element from sewage effluents is expensive and impractical. As a result, most antieutrophication programs have focused on phosphorus and relatively little time or effort has been expended on nitrogen.

SULFUR, POTASSIUM, CALCIUM, AND MAGNESIUM

Sulfur, potassium, calcium, and magnesium are the remaining macronutrients. Sulfur and potassium have been studied less in natural waters than any of the other macronutrients. It is probable that they are rarely, if ever, limiting to the growth of aquatic angiosperms, and their removal from sewage effluents has not seriously been considered as a means of reducing human-accelerated eutrophication.

Calcium and magnesium have received more attention, but not because of their direct role as essential plant nutrients. Rather, they have been studied because of the carbonate and bicarbonate anions associated with them. The range of calcium and magnesium concentrations in lakes and rivers is very wide and is directly associated with the concentrations of these elements in the soils and underlying rocks of the watershed. Waters in limestone regions will be very hard, with high levels of calcium and magnesium. Waters lying in watersheds having acid soils and parent materials will be soft, with low levels of these two elements.

Although certain plants and even certain plant communities have been identified as being "soft water plants," or "hard water plants" very little work has been done to separate the relative effects of the cations (calcium and magnesium), the anions associated with them (especially carbonate and bicarbonate), and the pH of the water. These factors are all intimately related, and it is possible that hard water plants do not require high concentrations of calcium as much as they require high pH levels. A great deal of additional research is necessary before we understand this situation completely.

MICRONUTRIENTS

Almost all of the studies on micronutrients in relation to aquatic plant growth have dealt with phytoplankton rather than higher plants. Goldman (1972) stated that trace element deficiencies are more likely to occur in oligotrophic lakes than in eutrophic lakes. Of 28 oligotrophic lakes he studied, 82% were found to be deficient in one or more trace elements. Even molybdenum, which is required in extremely minute amounts, was found to be limiting in one lake. It must be emphasized, however, that these studies were conducted using phytoplankton as the test organisms, and the findings should not be extrapolated to vascular plants without additional evidence.

The effects of summer stratification on iron have been elucidated in considerable detail because iron concentrations are closely associated with phosphorus concentrations. Ferric phosphate is very insoluble and under the reducing conditions found in the hypolimnion in late summer, iron precipitates out of solution in this form. Ferrous phosphate is considerably more soluble, and when oxygen is abundant, both iron (ferrous), and phosphorus ions return to solution. Manganese follows a similar pattern (Ruttner 1953).

Very little is known about the seasonal cycling or biological roles of boron, zinc, copper, or chlorine in natural aquatic environments. It is usually assumed that they are present in sufficient quantities in most waters and that plant growth is rarely limited by any of them. The primary factor controlling the supply of zinc, copper, and manganese is the pH of the soils in the watershed and of the water itself. These nutrients might possibly become limiting in high pH situations but never in low pH situations.

BIBLIOGRAPHY

EPSTEIN, E. 1972. Mineral Nutrition of Plants: Principles and Perspectives. John Wiley & Sons, New York.

FRINK, C.R. 1967. Nutrient budget: rational analysis of eutrophication in a Connecticut lake. Environ. Sci. Technol. 1, 425–428.

GOLDMAN, C.R. 1972. The role of minor nutrients in limiting the productivity of aquatic ecosystems. *In* Nutrients and Eutrophication. G.E. Likens (Editor). American Soc. Limnol. Oceanogr., Lawrence, KS.

HUTCHINSON, G.E. 1975. A Treatise on Limnology. Vol. III, Limnologial Botany. John Wiley & Sons, New York.

JONES, U.S.   1979.   Fertilizers and Soil Fertility. Reston Pub. Co., Reston, VA.

JUDAY, C., and BIRGE, E.A.   1931.   A second report on the phosphorus content of Wis. lake waters. Trans. Wisc. Acad. Sci., Arts, Lett. **26**, 353–382.

NICHOLAS, D.J.D.   1961.   Minor mineral elements.   : Annual Review of Plant Physiology. Vol. 12. Leonard Machlis (Editor). Annual Reviews, Inc., Palo Alto, CA.

NORVELL, W.A. and FRINK, C.R.   1975.   Water Chemistry and Fertility of Twenty-Three Connecticut Lakes. Conn. Agric. Expt. Sta., Bull. 759. New Haven, CT.

RUTTNER, F.   1953.   Fundamentals of Limnology. D.G. Frey and F.E.J. Fry. (Translators). Univ. of Toronto Press, Toronto, Ontario, Canada.

Part II

Biological Aspects
of Aquatic Plants

6

What Is an Aquatic Plant?

This is a book about aquatic vegetation but we have not yet decided which plants are to be included in that category. What is an aquatic plant? Plants in their natural habitats live under a very wide range of moisture conditions. At one extreme are members of the families Cactaceae and Mesembryanthemaceae. Certain species of the latter family (including curious plants known popularly as living stones and split rocks), live, and indeed thrive, in southwestern Africa, under conditions so dry that plant life seems impossible.

At the other extreme are members of various botanical families that complete their entire life cycles from seed to seed completely submerged beneath the surfaces of lakes, ponds, or streams. These plants never come into direct contact with the atmosphere and are completely surrounded by water throughout their lives. Many plants in this category will dessicate and die in a matter of minutes if they are removed from the water and placed in direct sunlight.

Between these two extremes lie the vast majority of vascular plant species. They require varying amounts of water, but they will tolerate neither extreme aridity nor prolonged submersion. Furthermore, these plants form a continuum between the two extremes. There are plants that will tolerate every conceivable level of moisture availability between the upper and lower limits described above.

Since plants do form a continuum and do not fall naturally into discrete categories in regard to their moisture requirements, the definition of "aquatic plant" is necessarily arbitrary. The question is, "Where should we draw the line?". Where in the continuous range from extreme xerophyte to extreme hydrophyte should we make an artificial and arbitrary break and say "everything beyond this point is an aquatic plant"? An even more vexing problem is trying to define this point in words after we have identified it to our own satisfaction in our own minds. Virtually every definition that has been written has some exceptions and some hazy, gray areas.

The problem is further confounded by the fact that some species of plants, while typically and normally found growing under certain moisture levels, can persevere and perhaps even grow, under other, very different, moisture levels which, for them, are highly abnormal. For example, some species of *Dracena* are regularly offered for sale in pet shops for use in freshwater aquaria. These are terrestrial plants that would never be found growing under water in nature. Nevertheless they persist for long periods of time completely submerged in home aquaria. This does not qualify them as aquatic plants!

Conversely, some plants, which are truly aquatic can survive for weeks or even months when their normally wet habitats dry out. A few species are even capable of producing so-called "land forms," which replace the typical submerged foliage when the habitat dries out. (See Chapter 9 for a more detailed discussion of this phenomenon.) These are still regarded as aquatic plants because the production of "land forms" is temporary and abnormal.

In the final analysis, it is not important that we have a strict definition of aquatic plant that never fails us and has absolutely no exceptions. It is far more important that we understand the requirements and tolerances of various species with which we are working and know where they fit into the continuum from extreme xerophyte to extreme hydrophyte. Whether a particular species happens to fall just to the left or just to the right of some arbitrary dividing line is really of little consequence.

Therefore, instead of a strict definition, the following general concept of the term aquatic plant is offered. An aquatic plant is a plant that is normally found in nature growing in association with free-standing water whose level is at or above the surface of the soil. The plant may be floating upon the water, completely submerged in the water, or partly submerged in the water. In some instances the plants may merely be growing near the water but in definite association with it. It must be remembered that individual plants found growing under these conditions do not qualify as aquatic plants if this is not their normal habitat.

TYPES OF VASCULAR PLANTS

Aquatic vascular plants can be categorized into four separate types based on their habit of growth. This classification has nothing to do with phylogenetic relationships; it is based solely on the way in which the plants grow in physical relationship to the water (Fig. 6.1). The four categories are floating unattached, floating attached, submersed, and emergent.

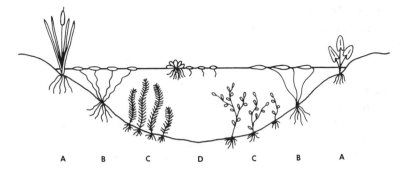

A   B   C   D   C   B   A

Fig. 6.1. Types of aquatic plants based on their growth habits.
(A) Emergent species. (B) Floating attached species. (C) Submersed
species. (D) Floating unattached species.

Floating Unattached Plants

These plants float with most of the plant body above the water's
surface. Roots, if present, hang free in the water and are not anchored to
the bottom. Floating unattached plants move about with winds and
currents. Examples of floating unattached plants are duckweed, water
hyacinth, and water lettuce. Some plants in this category are among the
world's worst weed problems.

Floating Attached Plants

Floating attached plants have their leaves floating on the water's surface but their roots are anchored in the substrate. The leaves are connected to the bottom by petioles alone (as in the water lilies), or by a
combination of petioles and stems (as in some pondweeds). Some floating attached plants have underwater leaves in addition to the floating
leaves.

Submersed Plants

Submersed plants are those that spend their entire life cycle, with the
possible exception of flowering, beneath the surface of the water. With
very few exceptions, they are anchored to the substrate, and the vegetative portion of the plant does not reach the surface or else the terminal end
lies in a horizontal position just beneath the surface. In many types the
flowers are borne above the water, but these are still classified as submersed species. Examples of submersed plants are watermilfoil, elodea,
many pondweeds, wild celery, and coontail.

Emergent Plants

Emergent aquatic plants are those whose roots and basal portions grow beneath the surface of shallow water but whose leaves and stems are borne primarily in the air. Examples are cattail, pickerelweed, some woody shrubs, and the many species of rushes and sedges found in shallow waters along lake margins and on stream banks.

With these concepts in mind, we are ready to move on to the next three chapters in which we will learn to identify some common aquatic plants, examine certain of their biological aspects in comparison to terrestrial species, and learn something of their ecological relationships.

7

Identification of Common Aquatic Plants

Proper identification of aquatic plants is a necessary prerequisite to a successful management program. One cannot assess the value of a marsh as waterfowl habitat without identifying the plants growing there. Neither can one plan a successful aquatic weed control program without proper identification. Aquatic species respond in different ways to different management practices, just as terrestrial plants do, and this necessitates positive identification.

Space in this book does not permit a key for the identification of aquatic plants. Instead, there are brief descriptions of various plant species, genera, and families with pictures of representative plants. The taxa that are included were chosen because of their ecological importance, their economic impact, or their conspicuousness in the field.

For more specific identification or identification of the many types of aquatic plants not included here, refer to the books listed in the bibliography at the end of this chapter. Most of them identify plants to species, although some such as Prescott (1969) and Cook (1974) identify them only to genus.

EMERGENT PLANTS

Arrow Arum (Genus *Peltandra*) Fig. 7.1

These are perennials with short, erect rootstocks and long-petioled, basal leaves with leaf blades shaped like arrowheads. Flowers are inconspicuous and are borne on a spadix enclosed within a spathe. Flower stalks re-curve, submerging the green, brown, or purplish berries beneath the surface of the water.

These plants can be distinguished from pickerelweed and arrowhead, even when they do not have flowers or fruits, by the single, straight, prominent vein that runs the length of each lobe in the leaf blade.

There are three species that occur in shallow waters of eastern North America.

Fig. 7.1.   Arrow Arum *(Peltandra virginica).* **(A) Plant showing leaves and spadix in spathe. (B) Mature fruits in spathe.** *Courtesy of N.J. Coop. Extension Service.* **(0.20×)**

Arrowhead (Genus *Sagittaria*) Fig. 7.2

This is an extremely variable genus of approximately 20 species, most of which occur in the new world. They have basal leaves whose blades are usually arrowhead-shaped, but which may vary from lance-linear to broadly ovate. White flowers are borne in 1 to 12 whorls, usually with three flowers per whorl. Rootstocks bear tubers, which are favorite foods of ducks and muskrats giving rise to one of the common names, duck potato.

When flowers or fruits are not present, this plant can be distinguished from arrow arum and pickerelweed by the presence of several slightly curved veins of equal prominence in each of the rearward-projecting lobes of the leaf blades.

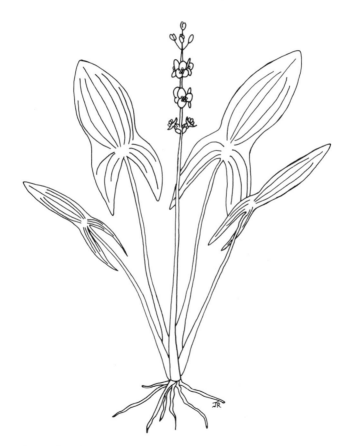

Fig. 7.2.   Arrowhead *(Sagittaria latifolia)*. Plant showing pattern of veins in leaves and flower in clusters of threes. (0.20×)

Bulrush (Genus *Scirpus*) Figs. 7.3 and 7.4

These are rushlike plants with stems triangular or round in cross section. The leaves have sheaths at their bases. Specialized leaves (called the involucre), arise from the base of the inflorescence and appear to be continuations of the stem. Identification to species is extremely difficult, and mature nutlets (achenes) are required. About 150 species of bulrushes are widely distributed around the world, but are most common in North America.

Fig. 7.3.   Bulrush *(Scirpus americanus)*. **(A) Upper portion of stem with spikelet cluster and involucral leaf. (B) Lower portion of stem and root-stock. (C) Cluster of spikelets.** *Courtesy of N.J. Coop. Extension Service.* **[(A) 0.5×, (B) 0.5×, (C) 1.5×]**

Fig. 7.4.   Great Bulrush *(Scirpus validus)*. (A) Upper portion of stem with involucral leaf and cluster of spikelets. (B) Lower portion of stems and rootstock. *Courtesy of N.J. Coop. Extension Service.* [(A) 0.5×, (B) 0.5×]

Bur Reed (Genus *Sparganium*) Fig. 7.5

This genus consists of erect perennial plants with alternate, ribbonlike leaves. They are weakly rooted with clustered, fibrous roots that allow the plants to be easily pulled up. The most distinguishing characteristics are the fruits, which are clustered together in tight, spherical green or brown "burs" from which the common name is derived.

There are about 20 species, mostly in the temperate and arctic portions of the Northern Hemisphere. All are capable of growing in the form of sterile rosettes when they occur in deeper water. These rosettes have flaccid leaves whose terminal ends bend over and lay in a horizontal position just beneath the water's surface.

Fig. 7.5.   Bur reed *(Sparganium americanum).* (A) Upper portion of stem with female and male heads. (B) Fruit. *Courtesy of N.J. Coop. Extension Service.* [(A) 0.5×, (B) 2×]

Cattail (Genus *Typha*) Fig. 7.6

Cattails are perennials with stout rhizomes and erect stems up to 300 cm (10 ft) tall. Leaves are flat, ribbonlike, and arise from near ground level. Pistillate portions of inflorescences persist for months and are the familiar brown, cigar-shaped "cattails" or "punks" which are gathered for home decoration or to burn for the pungent aroma they produce and for the smoke which is said to drive away mosquitoes. There are about 10 species worldwide, and many form extensive, monospecific stands in swamps, marshes, and along streams.

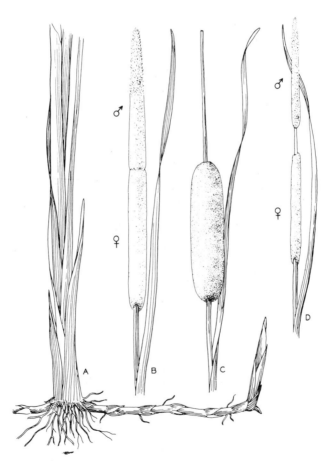

Fig. 7.6.   Cattail (*Typha* spp.). (A) *T. latifolia.* Lower portion of plant. (B)T. latifolia. Spike showing male and female flowers. (C) *T. latifolia.* Spike in fruit after male flowers have fallen off. (D) *T. angustifolia.* Spike in flower showing gap between male and female flowers. *Courtesy of N.J. Coop. Extension Service.* (0.25×)

Giant Reed (Genus *Phragmites*). Fig. 7.7

This is a tall, perennial grass with creeping rhizomes and/or stolons and hollow, round stems which may reach heights of 400 cm (13 ft). A distinctive characteristic is the large, feathery, plumelike panicle, which is 15−40 cm (6-16 in.) long. There are two or three species that have extremely wide distribution around the world, although they are less common in the tropics.

They are also known as "common reed," and they are most frequently found in fresh and brackish marshes and on the edges of lakes, ponds, and streams. They are occasionally found on dryer sites such as roadsides, particularly where rhizomes have been inadvertently introduced with fill.

It may be distinguished from wild rice by the panicle (see Fig. 7.7) and by the stout rhizomes that make giant reed difficult to pull up.

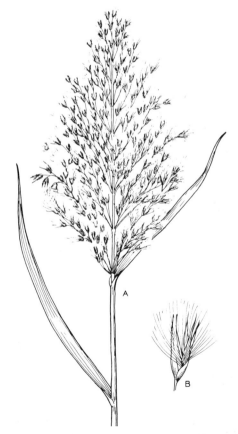

Fig. 7.7.   Giant reed *(Phragmites australis)*. (A) Upper portion of stem with inflorescence. (B) Spikelet with long hairs from base of flowers. *Courtesy of N.J. Coop. Extension Service.* **(0.25×)**

Pickerelweed (Genus *Pontederia*) Fig. 7.8

These are perennials with heart-shaped leaf blades growing on clump-ed, basal petioles. Blue or violet-blue, two-lipped flowers are borne on terminal spikes. Stems that bear flowers also bear a single leaf below the flower. Pickerelweed can be distinguished from arrow arum and arrow-head when flowers or fruits are not present by the numerous, small, curved veins in the two lobes of the heart-shaped leaf blade.

This plant frequently forms colonies in shallow, muddy areas along the margins of lakes, ponds, and streams. Five species occur, all in the new world.

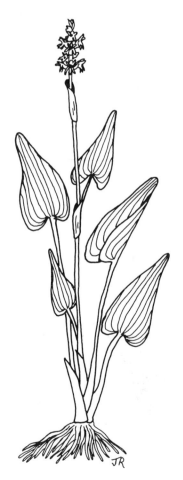

Fig. 7.8.   Pickerelweed *(Pontederia cordata)*. Whole plant showing pattern of veins in leaves and the single leaf growing on the flower stalk. (0.20×)

Rushes (Genus *Juncus*) Fig. 7.9

This is a very large genus of annual and perennial plants looking similar in appearance to grasses and sedges with which they are often confused. They are separated from the grasses by technical characteristics of the flowers and by the absence of a ligule at the junction of the leaf blade and the sheath.

Flowers are small and appear in terminal clusters called cymes. In some species the flowers appear to be on the side of the stem because an upper, erect leaf looks like a continuation of the stem (see Fig. 7.9). Stems are round in cross section and are pithy, hollow, or horizontally partitioned inside. A few species grow submerged. Most of the approximately 225 species occur in the Northern Hemisphere.

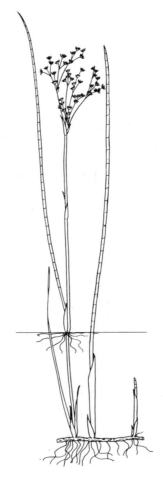

Fig. 7.9.   Bayonet rush *(Juncus militaris)*. Portion of plant with rootstock. *Courtesy of N.J. Coop. Extension Service.* **(0.15×)**

Sedges (Genus *Carex*) Fig. 7.10

The genus *Carex* is a very large (over 1000 species) and variable genus of superficially grasslike perennials. Inflorescences vary widely, but the fruit, an achene or nutlet, is always enclosed in a sac called the perigynium. Mature perigynia are required in order to identify sedges to species. Stems are usually solid and triangular in cross section.

Sedges occur in damp meadows, swampy areas, and shallow water throughout much of the world, but are most abundant in temperate zones.

Fig. 7.10.   Woolly sedge *(Carex lanuginosa)*. (A) Upper portion of stem with two female spikes (lower) and two male spikes (upper). (B) Perigynium. *Courtesy of N.J. Coop. Extension Service.* **(0.5×)**

Spikerush (Genus *Eleocharis*) Fig. 7.11

These are perennials (or rarely annuals) forming tufts or clumps of leafless, erect stems arising from subterranean rootstocks. Stems are round (or rarely four-angled) in cross section and contain numerous, nearly equal air canals. Flowers are borne in small, solitary, terminal spikelets. About 200 species are known, scattered over much of the world. Plants may grow completely submerged when they are in deeper waters, in which case they remain sterile.

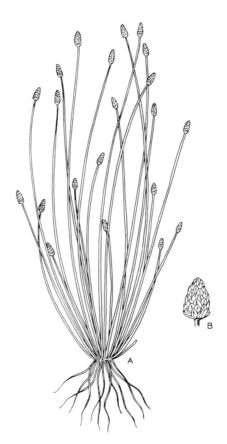

Fig. 7.11.   Spikerush *(Eleocharis obtusa)*. (A) Whole plant. (B) Fruiting spike. *Courtesy of N.J. Coop. Extension Service.* **(0.5×)**

Wild Rice (Genus *Zizania*) Fig. 7.12

Wild rice is an annual grass reaching heights of 300 cm (10 ft) or more and having large, flat, tapering leaves and a prominent panicle up to 50 cm (20 in.) long. They superficially resemble giant reed and grow in similar habitats, but can be distinguished by the panicle (see Figure 7.12), and by the ease with which wild rice is uprooted.

One or two species with several varieties occur in eastern North America, always on soft, muddy bottoms in areas of slow water circulation. They are highly prized as food for waterfowl and humans and have, therefore, been widely planted outside of their natural range.

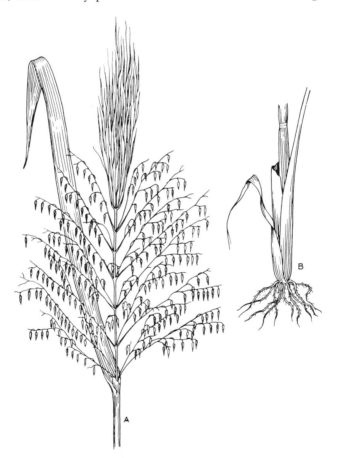

Fig. 7.12.   Wild rice *(Zizania aquatica)*. (A) Inflorescence with pendulous male flowers on lower branches and erect female flowers on upper branches. (B) Lower portion of stem. *Courtesy of N.J. Coop. Extension Service.* [(A) 0.20×, (B) 0.20×]

FLOATING ATTACHED PLANTS

Spatterdock (Genus *Nuphar*) Fig. 7.13

Plants of this genus are also known as yellow water lilies or cow lilies. Leaf blades are leathery, heart-shaped or nearly circular, with a deep cleft at the base and either float on the water's surface or protrude into the air, depending on the species. Petioles are stout and less flexible than in white water lilies. Flowers are large, yellow (at least on the inside), and often nearly spherical. They are borne above the water's surface (not floating), on a stout, rigid peduncle. Fruits ripen above the water. Rhizomes are enormous; up to 15 cm (6 in.) in diameter and 4.5 m (15 ft) in length. They are covered with semicircular leaf scars from previous season's leaves.

The number of species is probably between 7 and 25, but the taxonomy is poorly understood and many intergrades and hybrids occur.

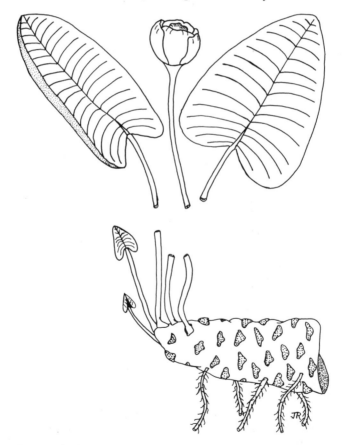

Fig. 7.13.   Spatterdock (Genus *Nuphar*). Two leaf blades, one flower, and terminal end of rhizome showing newly-emerging leaves and leaf scars from previous year's leaves. (0.15×)

Water Shield (Genus *Brasenia*) Fig. 7.14

Watershield has oval leaf blades up to 10 cm (4 in.) long, which float on the surface. They do not have a cleft at the base as do the water lilies, and the petiole is attached at the center of the blade. They are connected to the slender, creeping rhizomes by a system of branching stems and petioles. All underwater surfaces of the plant are covered with a thick, mucilaginous, colorless, jellylike material. Flowers are small and relatively inconspicuous, pale purple in color, and borne above the surface of the water.

There is only one species, and it is sporadically distributed in North and Central America, the West Indies, Africa, eastern Asia, and Australia.

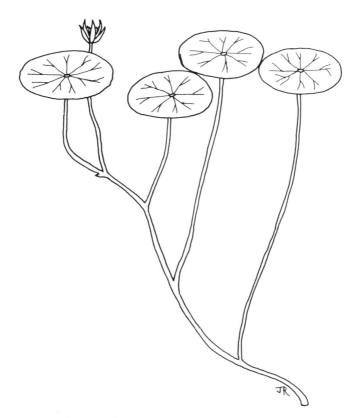

Fig. 7.14.   Water shield *(Brasenia schreberi)*. Upper portion of stem showing floating, peltate leaves and emergent flower. (0.5×)

White Water Lilies (Genus *Nymphaea*) Fig. 7.15

The leaf blades are circular or nearly circular, occasionally up to 40 cm (15 in.) in diameter, deeply cleft at the base, and float upon the surface of still water. Blades are attached to subterranean stems (either rhizomes or corms) by slender, flexible petioles. There is no stem above the surface of the substrate. Conspicuous flowers float on the surface, attached to the underground stems by slender peduncles. Flowers are usually white but may be pink, yellow, or blue. Fruits ripen beneath the water.

There are about 40 species, which are now cosmopolitan because of introductions made for ornamental purposes.

Fig. 7.15.   White water lily (Genus *Nymphaea*). Two floating leaves and flower arising from section of rhizome. (0.25×)

FLOATING UNATTACHED PLANTS

Duckweeds (Family *Lemnaceae*) Fig. 7.16

Duckweeds are the world's smallest flowering plants. They consist of minute flattened or globular bodies called fronds or thalli (singular: thallus), which float on the surface of quiet waters. The fronds rarely exceed 10 mm (0.4 in.) in any dimension, and in some genera they do not exceed 1 mm (0.04 in.). Tiny, simple, white roots may be present, depending on the genus. Distinct stems and leaves are not present, but tissues of both are thought to be incorporated into the thallus.

Thick mats made up of mixed genera and species are frequently found on still ponds and slowly moving waters. There are six genera with a total of approximately 30 species, which are cosmopolitan in distribution.

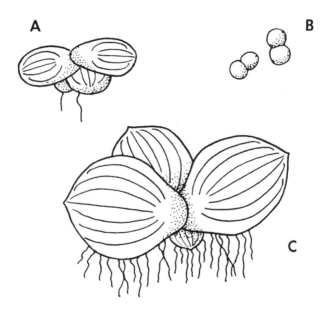

Fig. 7.16.   Three genera of duckweeds (Family *Lemnaceae*). (A)Lemna minor. Four plants in a cluster, each with a single root. (B) *Wolffia columbiana.* Four rootless plants. (C) *Spirodela polyrhiza.* Four plants in a cluster, each with numerous roots. *Courtesy of N.J. Coop. Extension Service.* (5×)

Water Fern (Genus *Salvinia*) Fig. 7.17

These are true ferns with horizontally floating branched stems and no roots. Leaves are in whorls of three at each node. Two leaves of each whorl are round, floating, and photosynthetic with white, erect, unwettable hairs on their upper surfaces and wettable hairs on their lower surfaces. The third leaf at each node is submerged, nonphotosynthetic, and divided into numerous hairlike segments. This leaf looks like, and may function as, a tuft of roots.

There are 12 species, all confined to the warmer parts of the world. They have become serious pests in parts of India, Ceylon, and Africa.

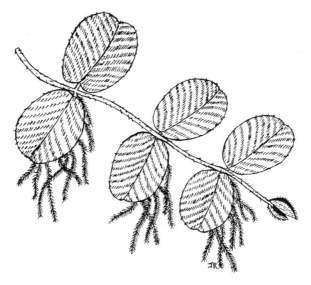

Fig. 7.17.   Water fern (Genus *Salvinia*). Horizontal floating stem with two floating leaves at each node along with one finely divided, submersed, rootlike leaf. (1.5×)

Water Hyacinth (*Eichhornia crassipes*) Fig. 7.18

The water hyacinth has been designated as one of the ten worst weeds in the world. Plants consist of floating rosettes of leaves with inflated, bladderlike petioles. Petioles are almost spherical when plants grow in uncrowded conditions but are elongated under crowded conditions. Roots are dark in color and fibrous, forming a dense mass beneath the rosette. Plants vary greatly in height from a few centimeters to over 1 meter (39 in.). Flowers are showy, being blue to purple with a yellow spot, and they are borne on a spike. Vegetative reproduction is common and plants may be connected by stolons. It is present in most tropical and subtropical parts of the world, having been introduced into many areas for its ornamental value.

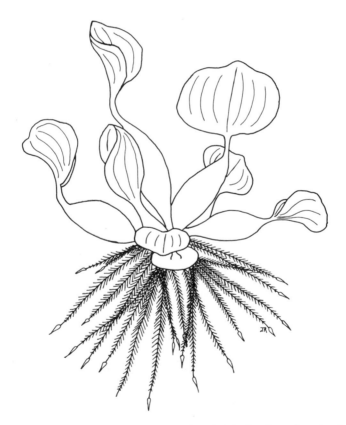

Fig. 7.18.   Water hyacinth *(Eichornia crassipes)*. Whole plant showing inflated, bladder-like petioles. (0.4×)

Water Lettuce (*Pistia stratiodes*) Fig. 7.19

Water lettuce plants consist of floating rosettes of pale, yellowish-green, fan-shaped leaves up to 15 cm (6 in.) long. They are deeply ribbed and covered with a soft pubescence, which is densest on the under side of the leaves. The flowers are small, green, and inconspicuous and are borne in the center of the plant among the leaf bases. Roots form a fibrous mass beneath the rosette. Reproduction is primarily or entirely vegetative by means of stolons. It is widely distributed throughout the tropics and subtropics and frequently becomes a pest.

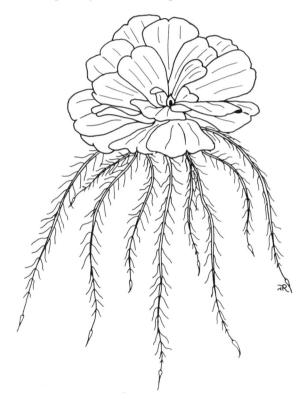

Fig. 7.19.   Water lettuce *(Pistia stratiodes)*. Whole plant with rosettelike cluster of leaves floating on surface. (0.33×)

Water Velvet (Genus *Azolla*) Fig. 7.20

These are tiny ferns with horizontally floating branched stems that are completely hidden by minute, overlapping, scale-like leaves that are less than 0.5 mm (0.02 in.) long. Leaves have two lobes: an upper, thick, green lobe, which does not touch the water, and a lower, thin, colorless lobe whose bottom surface is in contact with the water. Cavities in this surface are inhabited by a species of blue-green algae with the ability to fix nitrogen. Simple roots hang free in the water from the nodes. Older plants may become pink, red, or maroon, giving a pond or canal a unique appearance.

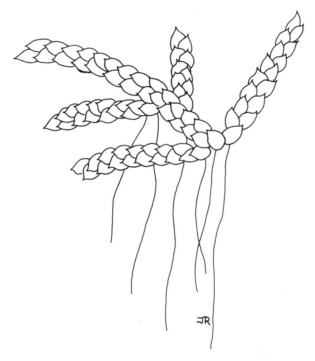

Fig. 7.20.   Water velvet (Genus *Azolla*). Whole plant showing overlapping, scalelike leaves and long, simple roots. (5×)

SUBMERSED PLANTS

Bladderwort (Genus *Utricularia*) Fig. 7.21

Bladderworts comprise a large genus with about 150 species, 25 or 30 of which are aquatic. These aquatic representatives have very flexible, thin stems that are totally devoid of roots and that are covered with hairlike simple leaves or compound leaves divided into hairlike segments. They frequently grow in a horizontal position, floating just beneath the surface of the water. Leaves bear tiny seedlike bladders with trap doors that serve as traps for small aquatic invertebrates. Bladders are clear or pale yellowish-green when empty, and dark colored when they contain animal remains.

Flowers are two-lipped, yellow or purple, with one or several borne on a thin, wiry scape that protrudes into the air. In some species the scape is supported by a whorl of inflated, spongy leaves that serve as pontoons and provide a stable platform to keep the flower above the water.

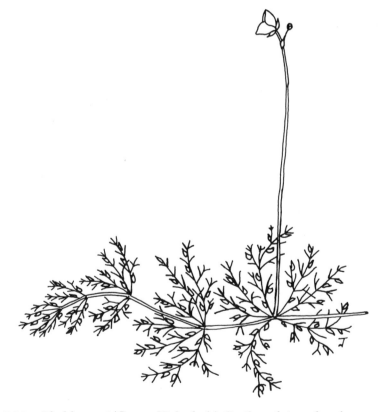

Fig. 7.21. Bladderwort (Genus *Utricularia*). Portion of stem showing emergent flower stalk and small bladders scattered among leaves. (0.66×)

Coontail (Genus *Ceratophyllum*) Fig. 7.22

Stems in this genus are long, thin, and flexible. Plants are entirely without roots even in the seedling stage. Leaves are flat and forked, being divided into several flat, narrow, stiff segments, somewhat like fanwort, but with fewer segments, and the segments are slightly curved. One species has minute teeth along one edge of each segment. Flowers and fruits are small and inconspicuous.

The degree of crowding of the whorls of leaves along the stem varies markedly. In some plants the whorls are crowded so tightly together that the stem is completely hidden and in other plants there are long sections of bare stem (internodes) between whorls. Whorls are frequently more crowded near the tip of the stem, giving the appearance of a raccoon's tail; hence the common name. The species are very difficult taxonomically and there may be anywhere from 2 to 30 species widely dispersed around the world, depending upon whose classification you use.

Fig. 7.22.   Coontail (Genus *Ceratophyllum*). Upper portion of stem showing whorled, forked leaves. *Courtesy of N.J. Coop. Extension Service.* **(0.5×)**

Elodea (Genus *Elodea*) Fig. 7.23

These perennials have slender, green, flexible stems that are moderately to sparsely branched and rooted to the bottom with fibrous roots. Leaves are without petioles and arise in whorls of three to seven along the entire length of the stem. (Leaves near the base of the stem may be alternate or opposite.) Leaves are strap-shaped, and, in the common American species, vary in width from 0.75 mm (0.03 in.) to 1.8 mm (0.07 in.). Minute white flowers rise to the surface on very long, white, threadlike stalks.

There are 17 species, originally confined to North and South (but not Central) America but now spread throughout the world due, at least in part, to its widespread use as an aquarium plant.

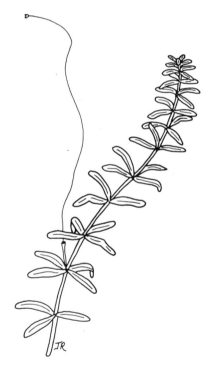

Fig. 7.23.   Waterweed *(Elodea canadensis)*. Portion of a stem showing whorled leaves and a female flower raised to surface on long, threadlike tube. (1×)

Fanwort (Genus *Cabomba*) Fig. 7.24

These are plants with long, thin, branched stems, which are weakly rooted to the bottom with fibrous roots. Submerged leaves are opposite or whorled, fanlike, and divided palmately into numerous, fine, flattened divisions. The floating white flowers are approximately 1 cm (0.4 in.) across and frequently have a yellow spot at the base of each petal. A tiny oval or forked floating leaf grows in association with each flower.

Some references mention a layer of mucilagenous, jellylike material on the leaves and stems as an identifying characteristic, but this is frequently absent. There are seven species in warm and temperate parts of the New World.

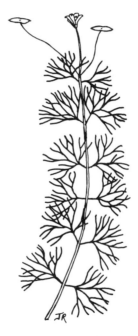

Fig. 7.24. Fanwort *(Cabomba caroliniana)*. Upper portion of stem showing finely divided underwater leaves, small peltate floating leaves, and emergent flower. (Compare palmately compound leaves of this plant with pinnately compound leaves of water milfoil, Fig. 7.31B.) (0.5×)

Hydrilla (*Hydrilla verticillata*) Fig. 7.25

This plant looks superficially identical to elodea (see Fig. 7.23). The differences are that hydrilla has slightly more prominent teeth along the leaf margins, forms underground tubers, and feels harsh, brittle, and coarse when drawn through the hand. Elodea has teeth on the leaf margins, but they almost require a hand lens to see them, it does not form underground tubers, and it feels soft and smooth when drawn through the hand.

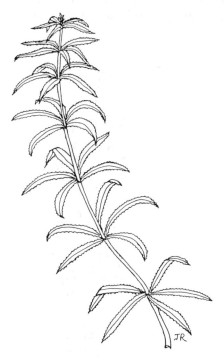

Fig. 7.25.   Hydrilla *(Hydrilla verticillata)*. Upper portion of stem showing tiny teeth along leaf margins. (1×)

Naiad (Genus *Najas*) Fig. 7.26

Plants in this genus closely resemble the pondweeds to which they are related. They differ from the pondweeds in having all the leaves opposite, although this may be obscured by the fact that there are bunches of smaller leaves in the axils of the larger leaves. All leaves are submersed, narrow, and ribbonlike. About 50 species are known which are cosmopolitan in distribution.

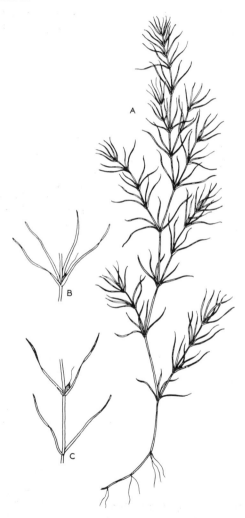

Fig. 7.26.   Naiad *(Najas flexilis)*. (A) Upper portion of plant. (B) Section of stem showing pseudo-whorled leaves and fruit. (C) Section of stem showing opposite leaves and fruit. *Courtesy of N.J. Coop. Extension Service.* **[(A) 0.5×, (B) 1×, (C) 1×]**

Pondweeds (Genus *Potamogeton*) 2 Figs. 7.27, 7.28, 7.29, and 7.30

This is a very large, variable, and complex genus whose taxonomy is poorly understood. Leaves are alternate (or rarely subopposite). Two basic types of leaves occur in the genus; floating leaves, which have a tough, leathery texture, and submersed leaves, which are thin, delicate, and frequently transluscent. Both types may occur on the same plant, although some species have only one or the other. Leaf shape varies from broadly ovate to narrow and hairlike. Petioles may be absent or present and stems are slender and branched. Stems are somewhat flexible but usually not flaccid.

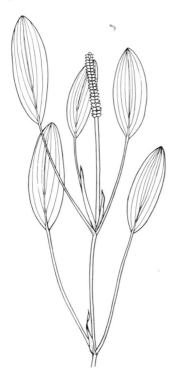

Fig. 7.27. Pondweed *(Potamogeton nodosus)*. Portion of plant showing broad, floating leaf blades, stipules in leaf axils, and stalked seed cluster. (Compare with Figs. 7.28, 7.29, and 7.30). *Courtesy of N.J. Coop. Extension Service.* (0.33×)

Fig. 7.28.   Pondweed *(Potamogeton robbinsii)*. Portion of plant showing submersed linear-lanceolate leaves and two stalked seed clusters. (Compare with Figs. 7.27, 7.29, and 7.30). *Courtesy of N.J. Coop. Extension Service.* (0.33×)

Flowers are small and crowded tightly into elongate or globose spikes that may be above or below the water's surface. A membranous stipule is present where the leaf joins the stem.

There are probably over 100 species worldwide, but there is much uncertainty regarding the taxonomy of individual species. Pondweeds are important foods for many wildlife species, and they are also serious weed problems. They are cosmopolitan in distribution and occur in an extremely wide variety of aquatic habitats.

Fig. 7.29.   Pondweed *(Potamogeton per-foliatus)*. Portion of plant showing leaves whose bases clasp the stem. (Compare with Figs. 7.27, 7.28, and 7.30). *Courtesy of N.J. Coop. Extension Service.* **(0.4×)**

Fig. 7.30.   Pondweed *(Potamogeton pusillus)*. Portion of stem showing narrow, threadlike leaves and small seed clusters. (Compare with Figs. 7.27, 7.28, and 7.29). *Courtesy of N.J. Coop. Extension Service.* **(0.5×)**

Watermilfoil (Genus *Myriophyllum*) Fig. 7.31

These plants have long, slender, flexible, sparingly branched stems with whorled or opposite leaves borne along their entire length. Leaves are divided into fine, threadlike segments, but, unlike fanwort or coontail, each leaf segment arises from a main central axis called a rachis. In some species the end of the stem becomes rigid and protrudes above the surface. The leaves of this emersed portion of stem are stiffer and have wider segments than the submersed leaves on the same plant. Flowers are tiny and inconspicuous and are borne in the leaf axils. There are about 40 species which are widely distributed but rare in Africa.

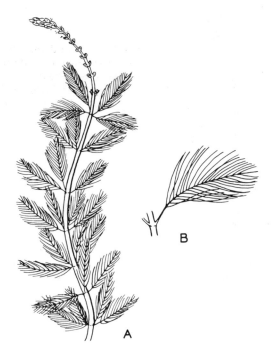

Fig. 7.31.   Water milfoil *(Myriophyllum spicatum)*. **(A) Upper portion of a flowering stem. (B) One pinnately compound leaf. (Compare with palmately compound leaf of fanwort, Fig. 7.24).** *Courtesy of N.J. Coop. Extension Service.* **(0.75×)**

Wild Celery (Genus *Vallisneria*) Fig. 7.32

Wild celery plants consist of clusters of long, flat, ribbon-like leaves arising from stolons or rhizomes. Leaves are up to 1 m (39 in.) in length, and the upper portions frequently float in a horizontal position just beneath the surface. Leaf tips are bluntly rounded and veins are clearly visible. Floating female flowers are elongate and tubular and are attached to the base of the plant by an extremely long peduncle that coils after pollination, retracting the developing fruit beneath the surface.

There are 6 to 10 species,which are almost cosmopolitan but absent from the colder regions. They are widely used as aquarium plants and are an excellent source of waterfowl food.

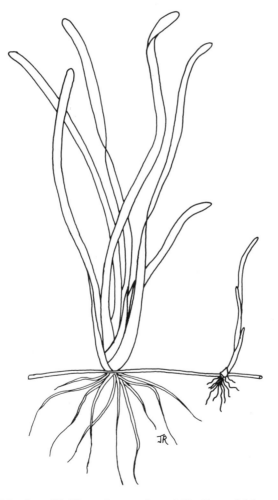

Fig. 7.32.   Wild celery *(Vallisneria americana)*. Portion of rhizome with mature plant at left and developing plant at right. (0.25×)

BIBLIOGRAPHY

ASTON, H.I. 1973. Aquatic Plants of Australia. Melbourne University Press, Melbourne, Australia.

CORRELL, D.S., and CORRELL, H.B. 1975. Aquatic and Wetland Plants of Southwestern United States. (Two volumes). Stanford University Press, Stanford, CA.

COOK, C.D.K. 1974. Water Plants of the World. Dr. W. Junk b.v., Publishers, The Hague, Netherlands.

CROW, G.E., and HELLQUIST, C.B. 1981. Aquatic Vascular Plants of New England: Part 2. Typhaceae and Sparganiaceae. Station Bull. No.517, New Hampshire Agricultural Experiment Station, Durham, NH.

CROW, G.E., and HELLQUIST, C.B. 1982. Aquatic Vascular Plants of New England: Part 4. Juncaginaceae, Scheuchzeriaceae, Butomaceae, Hydrocharitaceae. Station Bull. No. 520, New Hampshire Agricultural Experiment Station, Durham, NH.

FAIRBROTHERS, D.E., MOUL, E.T., ESSBACH, A.R., RIEMER, D.N., and SCHALLOCK, D.A. 1965. Aquatic Vegetation of New Jersey. Extension Bull. No. 382, Extension Service, College of Agriculture, Rutgers University, New Brunswick, NJ.

FASSETT, N.C. 1966. A Manual of Aquatic Plants. University of Wisconsin Press, Madison, WI.

GODFREY, R.K., and WOOTEN, J.W. 1979. Aquatic and Wetland Plants of Southeastern United States. The University of Georgia Press, Athens, GA.

HELLQUIST, C.B., and CROW, G.E. 1980. Aquatic Vascular Plants of New England: Part 1. Zosteraceae, Potamogetonaceae, Zannichelliaceae, Najadaceae. Station Bull. No. 515, New Hampshire Agricultural Experiment Station, Durham, NH.

HELLQUIST, C.B., and CROW, G.E. 1981. Aquatic Vascular Plants of New England: Part 3. Alismataceae. Station Bull. No. 518. New Hampshire Agricultural Experiment Station, Durham, NH.

HELLQUIST, C.B., and CROW, G.E. 1982. Aquatic Vascular Plants of New England: Part 5. Araceae, Lemnaceae, Xyridaceae, Eriocaulaceae, and Pontederiaceae. Station Bull. No. 523. New Hampshire Agricultural Experiment Station, Durham, NH.

HOTCHKISS, N. 1972. Common Marsh, Underwater, and Floating-Leaved Aquatic Plants of the United States and Canada. Dover Publications, New York.

MASON, H.L. 1957. A Flora of the Marshes of California. University of California Press, Berkeley, CA.

MUENSCHER, W.C. 1944. Aquatic Plants of the United States. Comstock Publishers, Ithaca, NY.

PRESCOTT, G.W. 1969. How to Know the Aquatic Plants. W.C. Brown Co., Dubuque, IA.

STEWARD, A.N., DENNIS, L.J., and GILKEY, H.M. 1963. Aquatic Plants of the Pacific Northwest. Oregon State University Press, Corvallis, OR.

WELDON, L.W., BLACKBURN, R.D., and HARRISON, S.D. 1973. Common Aquatic Weeds. Dover Publications, New York.

Ecological Relationships Involving Aquatic Plants

Aquatic plants are involved in numerous ecological relationships with their physical and chemical environments, with each other, and with animals sharing their environments. These relationships are exceedingly complex and although they are discussed below as separate, individual topics, it should be remembered that in reality many of them are interrelated and form a web of relationships, each of which influences many others, directly or indirectly.

Thousands of such relationships could be discussed but only a few of the more important ones are included here as examples. The list is not intended to be complete by any means.

FOOD RELATIONSHIPS

All animals are dependent on green plants for their food, and fish, aquatic insects, and other aquatic animals are not exceptions. Most bodies of water are semi-closed systems with regard to food materials; that is, a high percentage of the food consumed by animals in most bodies of water is produced in that body of water. There is, of course, some exchange of food materials between the water and the surrounding terrestrial ecosystems. Dragon fly nymphs, for example, live in the water and utilize food from the aquatic system. When they mature and leave the water, they may be eaten by birds, which represents a contribution of the aquatic system to the terrestrial system.

Conversely, a grasshopper may feed all of its life on land and then fall into the water and be eaten by a fish. This represents a contribution of the terrestrial system to the aquatic system. For the most part, however, these exchanges are minimal, and the food utilized in an aquatic system is food that was produced by photosynthesis in the same system.

Natural food chains are inefficient in transferring energy from one trophic level to the next. Because of this inefficiency, large amounts of plant material are required to produce relatively small amounts of fish. This relationship of a large weight of primary producers relative to an ever-decreasing weight of consumers with each step upward in trophic level, is often depicted as a food pyramid. Such a pyramid is shown in Fig. 8.1.

Food Chains and Fish

Increasing primary productivity by increasing the growth of plants in a body of water will increase fish production, but all plants are not equally efficient in contributing to the food chain. For numerous reasons, phytoplankton is more desireable than either filamentous algae or vascular plants for purposes of increasing fish production—particularly warm water pond fish such as bluegill sunfish and large mouth bass. It has been known for many years that the higher plants do not contribute substantially to aquatic food chains. Over 60 years ago Shelford (1918) said "one could probably remove all the larger plants and substitute glass structures of the same form and surface texture without greatly affecting the immediate food relations." Thienemann (1935) said "The reed-rush

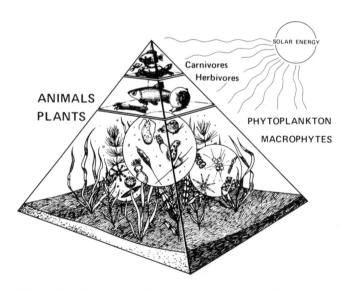

Fig. 8.1.   Diagrammatic representation of a food pyramid in a freshwater lake. *Courtesy of Wisconsin Coop. Extension Programs.*

zone serves no animal directly as nourishment, if one ignores certain mining insect larvae."

These statements are not completely accurate because some fish and other primary consumers do feed directly on aquatic vascular plants. Several such animals are discussed in some detail in Chapter 11. The principle is still valid, however, that phytoplankton are usually more effective bases for food pyramids than are higher plants, when fish production is the ultimate goal.

One reason for this is that phytoplankton is able to replenish itself very quickly after being grazed. The short life cycle of most phytoplankters along with their ability to absorb and assimilate nutrients very rapidly, assures a constant standing crop as long as conditions are favorable for growth and nutrients are available. Vascular plants might require days, weeks, or even longer to replenish themselves, whereas planktonic algae can replenish itself in a matter of hours under ideal conditions.

A second reason that phytoplankton is more efficient is that it has a higher nutritional value, pound for pound, than most vascular plants. This is due, in part at least, to the fact that plankters do not have conductive tissues or supportive tissues composed primarily of cellulose and therefore of little nutritional value to fish. It should be noted, however, that the nutritive content of any species may vary widely, depending on the conditions under which the plants were grown, their age, and possibly other factors.

A third reason is that the small size of plankters makes them suitable as food for a much wider variety of organisms in the food chain. Many primary consumers, including some of considerable size such as freshwater mussels, are adapted to feeding on plankton. Relatively few, particularly invertebrate organisms, are adapted to feeding on macroscopic plants. Submerged macroscopic plants in particular, are fed upon very little by invertebrates.

Another problem with rooted, vascular plants is that they interfere with the desirable predation on young fish. This sounds like a contradiction. If fish production is the primary objective in managing a particular body of water, why would predation on young fish be desirable? The answer lies in the fact that during each breeding season, far more young fish are produced in any given body of water than that body of water can support.

A pond or lake can be looked upon as an aquatic pasture that produces enough food to support only a given weight of fish. This weight may be present in the form of small numbers of large fish or large numbers of small fish, but the total carrying capacity of the pond cannot be exceeded for any appreciable length of time.

The growth of fish is dependent more on food supply than on age. Unlike mammals, fish can be held at almost any size by providing them with just enough food to keep them alive and healthy but with no excess calories for growth. Such fish are said to be stunted. They are small, but they may be perfectly healthy, and they are capable of becoming sexually mature and reproducing.

If the enormous numbers of newly hatched fish are not reduced by predation each year, they grow until they collectively use the the entire available food supply each day. At that point they stop growing but do not starve to death. They merely continue to exist at a nonharvestable size and continue to utilize the entire food supply. Such stunted, nonharvestable populations are undesireable from a human point of view. The fish are too small to utilize and no new growth is able to occur either in individual fish or in the population as a whole.

If sufficient predation occurs, the number of individual fish is reduced and the average size of the individual fish increases. The smaller fish continue to grow, since their collective demand for food is reduced along with their numbers. The larger fish continue to grow because they are feeding on the small fish. Predation is therefore necessary to maintain a growing, vigorous, fish population, especially in smaller bodies of water.

A food chain based primarily on macroscopic plants requires such dense stands that the plants interfere with this predation. Small fish are able to hide among them and escape detection. In extreme cases the plants can even hinder the movements of the larger fish and physically interfere with their catching of their prey. In a similar way, dense stands of vascular plants harbor snails, leeches, aquatic insects, crayfish, and other fish-food organisms. These invertebrates are not available to fish and thus the nutrients contained in them do not contribute to fish production.

A final reason that microscopic plants are more desirable as primary producers in fish production ponds than vascular plants is that the phytoplankters decompose more quickly and more completely when they die than vascular plants do. The vascular elements and supportive tissues in the higher plants have high carbon-to-nitrogen ratios and leave carbonaceous residues, which contribute to the filling-in of the pond and may also tie up other nutrient elements in a form not available for new plant growth.

Food Chains and Waterfowl

Aquatic plants form a high percentage of the diet of many species of waterfowl. The diet of the green-winged teal, for instance, consists of 100% plant material during the summer months, 91% plant material during the winter months, and 86% plant material during the spring and fall months. In contrast, the bufflehead's diet consists of 25% plant material in winter, 24% plant material in fall, 19% plant material in spring, and only 13% plant material during the summer months (Martin *et al.* *1961*).

All aquatic plants are not equally useful as waterfowl foods. Some, such as the pondweeds, wild rice, and wild celery are relished by ducks and geese and are heavily utilized by them. Others, such as watermilfoil, and fanwort are almost universally ignored by them. Thus the mere presence of aquatic plants does not make a body of water attractive to waterfowl; they must be the proper species of plants. A great deal of waterfowl management consists of maintaining the proper species of plants in the marshes and other bodies of water being managed.

In the case of some plants, such as duckweeds, the entire plant is eaten. In the case of others, such as wild rice, only the seeds are eaten and in still others, tubers or other specialized organs are consumed.

OXYGEN RELATIONSHIPS

Oxygen Production during Photosynthesis

The relative importance of photosynthetically produced oxygen compared to dissolved atmospheric oxygen in aquatic ecosystems is dependent to a great extent on the degree of mixing that occurs in the water. Although dissolution of oxygen from the atmosphere occurs at the surface, its downward movement into a body of water requires mechanical mixing or mass flow. The rate of diffusion of dissolved oxygen through water is so slow that it is of no practical importance in transporting this element to the deeper strata of lakes or ponds. In still waters with little mixing, therefore, photosynthetic oxygen may be of significant value in maintaining animal life at greater depths than would otherwise be possible. At the greatest depths, where oxygen is most likely to be limiting, light is often below the compensation level and photosynthetic plants cannot exist anyway.

Oxygen Reduction by Aquatic Plants

Many people are aware of the oxygen-generating capacity of plants and associate their presence with well-oxygenated waters. Few people are aware that aquatic plants can also reduce dissolved oxygen concentrations under certain conditions.

Every time the light intensity falls below the compensation level, respiration exceeds photosynthesis and plants consume more oxygen than they produce. Every night and on some dark days this happens and when several dark days occur in a row, the oxygen supply in a weedy pond can be depleted. Algae are more frequently implicated than are higher plants in such plant-induced oxygen deficiencies but higher plants can have the same effect.

A solid cover of free-floating plants such as water lettuce or duckweed can also reduce dissolved oxygen levels. This occurs because the plants shade the water column below them and prevent photosynthesis by submerged species. At the same time, the floating plants reduce the area of contact between the water and the atmosphere, thus reducing the dissolution of oxygen from that source.

In addition, some floating species continually shed dead roots, leaves and other organic debris into the water, which adds to the biochemical oxygen demand through their decay. The amount of organic matter shed into the water can be very substantial. Timmer and Weldon (1967) reported an accumulation of 17.8 cm (7 in.) of partially decayed plant material under a solid cover of water hyacinth in a period of approximately 6 months. They estimated that the oxygen demand of the daily pollutional load imposed on a body of water by one acre of water hyacinth is equal to that of the domestic sewage generated by 40 people.

A final means by which aquatic plants can reduce dissolved oxygen concentrations is the decay of large volumes of vegetation following a mass death of the plants. This may be a natural die-off precipitated by weather, disease, or other causes, or it may be a result of an intentional herbicide application. In either case the end result is decaying organic matter and depressed oxygen levels.

pH RELATIONSHIPS

The relationship between water pH and the distribution of aquatic plants is complicated and not well understood. As pointed out in Chapter 5, it is difficult to separate the effects of pH, calcium and magnesium, and carbonates and bicarbonates on submerged plants. The effects of the plants on water pH, however, is well understood.

As explained in Chapter 5, CO_2 forms a complex equilibrium when dissolved in natural waters (see Eq. 5.1). During active photosynthesis, free CO_2 is removed from the water (left side of the equation), and this causes the equilibrium to shift to the left, resulting in less carbonic acid in the water and a rise in the pH. When respiration exceeds photosynthesis, free CO_2 is added to the water (left side of the equation), the equilibrium shifts to the right adding carbonic acid to the system, which lowers the pH. Aquatic plants therefore tend to lower the pH of a body of water during the night and raise it during the day.

The actual fluctuation in pH depends on the concentration of plants and the buffering capacity of the water. In poorly buffered waters (generally those with low levels of carbonates, bicarbonates, and phosphates), aquatic plants can cause a pH variation of 3 full units between sunrise and sunset. For example, a soft-water pond may have a pH of 6.5 at sunrise and 9.5 at sunset. Such extreme variation in a short period of time might easily affect other organisms in the pond and perhaps even limit the species of other organisms which exist there.

SURFACE AREA RELATIONSHIPS

Many organisms, both plant and animal, live attached to, or in association with, a solid surface. Submerged foliage and stems of aquatic plants provide large surface areas that are suitable for colonization by such organisms. Several conflicting and overlapping terms have been used to describe these surface-associated communities. Epiphytes are sessile plants growing on the surfaces of other plants but not parasitic on them. The leaves and stems of many submersed vascular plants are covered with epiphytic algae.

Periphyte is a somewhat less well-defined term that always refers to organisms living on the surfaces of submerged plants. Some authors have used the term to mean aquatic epiphyte and others have used it in a broader sense.

Aufwuchs is a German term that is much broader and more inclusive than epiphyton. In addition to attached organisms, aufwuchs includes such free-living forms as leeches, and snails, and some insects and protozoans that creep on, walk on, or cling to solid surfaces. Many of these organisms, such as dipteran larvae of the family Chironomidae, are important links in the food chains between green plants and fish. The presence of vascular aquatic plants increases the population of aufwuchs in two distinctly different ways.

First, the presence of the plants enormously increases the inhabitable surface area beyond what which would be available on the unvegetated substrate alone. Edwards and Owens (1965) used a complex photographic technique to determine that a wide variety of dissected-leaf aquatic plants have a surface area of 1 cm^2 for each mg of dry weight of top growth. Using this value, they calculated that the ratio of plant surface area to substrate surface area ranges between 30:1 and 50:1 in fertile streams during the summer growing season. This means that the potential area for colonization by aufwuchs on the plants is 3000% to 5000% larger than the substrate upon which the plants are growing.

The increased surface area afforded by the plants is not the only means by which they increase the aufwuchs, however. The animal populations living on submerged plant surfaces are 300%−400% greater than those living on a silt bottom of the same surface area. On bare rock or gravel the difference is even more pronounced. Animal populations are as much as 1500% greater per unit area on plant surfaces than on rock or gravel substrates (Needham 1938). The increase in animals per unit area coupled with the increased area available, results in enormous increases in aufwuchs in vegetated areas as compared to nonvegetated areas.

All species of submerged plants are not equally populated by invertebrate animals. Table 8.1 presents the average numbers of individual organisms and the average numbers of genera found on seven different species of submersed aquatic plants in Lake Erie. Conditions within a solid stand of water milfoil might be quite different from conditions within a solid stand of wild celery. In order to minimize the impact of these differences on the results of the study, all plants were taken from mixed beds of vegetation so that the individual plants were growing under as nearly identical environmental conditions as possible. It is obvious from this table that some species of plants harbor much larger invertebrate populations than others, but the reasons for this are not clear.

It is possible that the difference is due, at least in part, to the degree of protection from predation that the plant affords to the invertebrates. *Myriophyllum spicatum* has a bushy habit of growth that provides excellent hiding places for the animals living on its surface. It was found to host the densest population of invertebrates. *Vallisneria spiralus* with its flat, ribbonlike leaves provides few hiding places, and it was found to have the least dense population of invertebrates. It is possible that predation is a factor in controlling the population density and that without predators, the numbers would be more nearly the same for both species.

It should now be clear why Shelford's statement, quoted earlier in this chapter, specified that if the higher plants were removed, they would have to be replaced by "glass structures of the same form and surface

TABLE 8.1. Invertebrate Organisms Found on Seven Species of Submersed Aquatic Plants in Lake Erie

Plant Species	Avg. No. of Organisms/10 ft of Stem	Avg. No. of Genera/Plant
Myriophyllum spicatum	1442	22
Potamogeton crispus	1139	22
Potamogeton compressus	572	10
Elodea canadensis	564	26
Potamogeton pectinatus	469	14
Najas flexilis	381	24
Vallisneria spiralus	30	4

From Krecker (1939).

texture" in order to not affect the immediate food relations. These glass structures would presumably provide the surfaces required for the auf-wuchs that are so vital in the food chain.

FILLING IN OF LAKES AND PONDS

Dense stands of vascular aquatic plants accelerate the rate of filling-in of standing bodies of water in three distinct ways. The remains of dead plants settle to the bottom, and, in many instances, they never decompose completely because the decomposing microorganisms run out of nitrogen before they are able to decompose all of the carbon-rich cellulose and related compounds. These carbonaceous remains continue to accumulate year after year and contribute significantly to the conversion of the lake or pond into a swamp, and ultimately to dry land.

Plants contribute to the filling-in process in a second, very different way. All flowing waters carry a certain amount of silt, clay, and organic detritus in suspension. When streams enter the relatively quiet waters of a lake or pond, much of the silt load drops out of suspension, however, a certain percentage will stay in suspension and wash right through the body of water and out of the outlet. Dense populations of plants in the lake or pond slow down the current and cause the water to take a longer, more circuitous route to the outlet. This causes more silt and debris to drop out of suspension and accelerates the rate of filling in, hence hastening the demise of the lake or pond.

The final method by which plants contribute to the filling-in process involves the electrical charges on the surfaces of clay particles and on the surfaces of the plants. Clay particles contain negative charges on their surfaces that function to keep the particles dispersed and in suspension. Some aquatic plants have been shown to have positive charges on their surfaces that attract the oppositely charged clay particles and cause them to form aggregates that are larger and heavier than the individual particles, and thus drop out of suspension more readily. This phenomenon has not been studied in detail and the amount of sediment actually deposited in natural bodies of water by this means is unknown.

By increasing the rate of siltation, aquatic plants alter their own environment. As bodies of water get shallower and shallower, submerged species in the deeper portions of the lake or pond may be replaced by attached, floating-leaved species that encroach further and further toward the center of the body of water. The emergent and shoreline species all extend further and further out from the original shoreline, and eventually there is no open water left at all. Thus, a succession of plant communities occurs over a period of time at any one point in the lake, and the plants themselves contribute to the rate at which the succession occurs by hastening the siltation process.

INCREASED WATER LOSS

Aquatic plants with aerial leaves serve as biological pumps, "pumping" water vapor into the air through the process of transpiration. The primary environmental parameters controlling transpirational loss of water are temperature, air movement, and relative humidity. Despite the high relative humidity that exists immediately above the surface of a body of water, transpiration from emergent foliage may be considerable. Timmer and Weldon (1967) found that a solid cover of water hyacinth increased the loss of water from a small pond by 370% over the loss from an open water surface under the same environmental conditions. They also found that the amount of water lost correlated very closely with total solar irradiation. Insolation would be expected to increase transpiration since it would raise the temperature, lower relative humidity, and might increase air movement by causing convective currents.

Aquatic plants can also increase water loss in a minor way by increasing the surface area from which evaporation can occur. A mat of sphagnum moss around the margin of a pond, for example, can serve as a giant wick, removing water from the pond and allowing it to evaporate from the plant surfaces.

NITROGEN FIXATION

We have seen in Chapter 5 that elemental nitrogen as the diatomic gas cannot be utilized by plants until it has been fixed in the form of nitrate or ammonia. Various species of bacteria and blue-green algae are capable of fixing nitrogen, but none of the higher plants have this ability. One genus of aquatic ferns, however, has developed a unique symbiotic relationship with a blue-green alga that is, in many ways, comparable to the relationship between legumes and *Rhizobium*. *Azolla* has small pouches on the lower surfaces of the fronds that are inhabited by the nitrogen fixing alga *Anabaena azollae*. The fern provides nutrients and protection for the algae, and the algae provides fixed nitrogen for the fern.

Azolla is a remarkably prolific plant. Under suitable conditions it can double its weight every 3–5 days and fix nitrogen at a rate that exceeds that of the legume/*Rhizobium* association. It is capable of fixing 2–4 kg of nitrogen/hectare/day (1.78 to 3.57 lb/acre/day) (Lumpkin and Plucknett 1982). This is a very substantial amount of fixed nitrogen and has a profound effect on the nutritional status of the water in which the plants live. The nitrogen becomes available to other plants on the death and decomposition of the fern.

Azolla has been cultivated for centuries in China and other Asian countries where it is used as a green manure in rice-growing areas. Exacting and often complicated techniques have evolved for managing the Azolla "crop." It is carefully tended, overwintered in special areas, and introduced to the rice paddies at the proper time. Some work has even been done on methods of controlling *Azolla* pests such as snails, insects, and fungi. An excellent review of the use of *Azolla* as a nitrogen source for crops is presented in the book by Lumpkin and Plucknett (1982).

BIBLIOGRAPHY

EDWARDS, R.W., and OWENS, M. 1965. The oxygen balance of streams. *In* Ecology and the Industrial Society. G.T. Goodman, R.W. Edwards, and J.M. Lambert (Editors). Symp. British Ecological Soc. **5**, 149−172. Blackwell, Oxford. *Cited by* C.D. Sculthorpe. The Biology of Aquatic Vascular Plants, St. Martin's Press, New York.

KRECKER, F.H. 1939. A comparative study of the animal populations of certain submerged aquatic plants. Ecology **20**, 553−562.

LUMPKIN, T.A., and PLUCKNETT, D.L. 1982. Azolla as a Green Manure: Use and Management in Crop Production. Westview Press, Boulder, CO.

MARTIN, A.C., ZIM, H.S., and NELSON, A.L. 1961 American Wildlife and Plants: A Guide to Wildlife Food Habits. Dover Publications, New York.

NEEDHAM, P.R. 1938. Trout Streams. Comstock Pub. Co., Ithaca, NY. *Cited by* C.D. Sculthorpe. The Biology of Vascular Plants. St. Martin's Press, New York.

SHELFORD, V.E. 1918. Conditions of existence. *In* Fresh Water Biology, H.B. Ward and G.C. Whipple. John Wiley & Sons, New York.

THIENEMANN, A. 1935. Die Bedeutung der Limnology fur die Kultur der Gegenwart. Stuttgart. *Cited by* W.C. Frohne. Limnological role of higher aquatic plants. Trans. Am. Micros. Soc. **57**, 256−268.

TIMMER, C.E., and WELDON, L.W. 1967. Evapotranspiration and pollution of water by water hyacinth. Hyacinth Control J. **6**, 34−37.

9

Adaptions of Aquatic Plants

In general, aquatic plants do not form large taxanomic groupings of their own that are set apart from other groups merely because they are aquatic. Many aquatic species belong to families and even genera that also include terrestrial representatives. In the United States, for example, there are approximately 1300 species of aquatic plants that are assigned to 306 genera in 65 different families. Of these 65 families, only 12 are wholly aquatic and the remaining 53 contain terrestrial species in addition to their aquatic representatives (Prescott 1980). This is an indication that there are no basic phylogenetic differences between aquatic and terrestrial plants. Two species may be closely related despite the fact that one lives in water and the other lives on land.

It is generally agreed that flowering plants evolved from primitive algae or algaelike ancestors, but that they evolved on land, not in the water. Today's aquatic spermatophytes are specialized forms that have reinvaded aquatic environments rather than being primitive forms that never left the water. In this respect they are much like whales and other marine mammals that have reinvaded the sea, although their ancestors were terrestrial.

Since aquatic spermatophytes are closely related to and descended from terrestrial spermatophytes, it is not surprising that the morphology and anatomy of the two groups are basically the same. Those differences that do occur are primarily adaptations to the environments in which the aquatic forms live. The degree of adaptation varies from slightly modified plants to highly specialized forms that are completely dependent on a submerged existence.

In addition to their modified characteristics, most aquatic plants (particularly submersed types), exhibit characteristics that were inherited from terrestrial ancestors and that are not well adapted to life underwater. This is analogous to the fact that whales have retained their lungs rather than reverting to gills or some other system for obtaining oxygen from the water in which they live.

In this chapter we will examine some of the adaptations exhibited by hydrophytes and we will examine some of the evolutionary solutions to problems presented by the inheritance of terrestrial characteristcs.

REPRODUCTION

Sexual Reproduction

The production of pollen rather than free-swimming sperm cells is a definite impediment in the underwater environment (De Wit 1978). Flowers of most aquatic plants, including submersed species, must be elevated above the water in order for pollination to occur by insects (entomophily), or wind (anemophily). Getting the flower above the surface and keeping it there, at least until pollination occurs, requires special, and sometimes highly elaborate, adaptation. In a few species, flowers are borne beneath the surface and transfer of pollen takes place under water (hydrophily). This is relatively rare, but it also requires some special adaptations.

As an example of a highly specialized and complex type of flowering and pollination we will use *Vallisneria spiralis*, a submerged plant with long, flat, tapelike leaves. The female flower is in the form of an elongate cylinder. It forms near the base of the plant, but as it nears maturity, the peduncle grows very rapidly (up to 2 cm/hr), by cell elongation, which allows the flower to float in a horizontal position on the water's surface, in a little depression formed by the surface tension. The peduncle is longer than the minimum necessary to just reach the surface, which allows some accomodation for sudden changes in water level. The sepals at the apical end of the flower spread apart after the flower reaches the surface. This exposes the stigmas, which are covered with tiny dense hairs that trap a layer of air, keeping the stigmas dry if they are temporarily submerged by a wave.

The male flowers are borne in a tubular spathe at the base of the plant. Unlike the female flowers, the peduncle does not elongate prior to maturity. Instead, the entire spathe abscises and floats to the surface completely free of the parent plant. Only then do the three tepals at the end of the spathe open, exposing the stamens. The tepals recurve, the larger two entering the water and serving as rudders and the smaller one remaining in the air and functioning as a sail. In this manner the male flowers are blown about on the surface with the stamens protruding.

When a male flower is blown close enough to a female flower, it slides down the slope into the depression caused by the weight of the female flower and the tips of the stamens touch the stigmas, transferring the pollen. Following pollination the peduncle forms into a coil, pulling the flower near to the bottom where the fruits develop and the seeds mature. Presumably this reduces the chances of the fruits being eaten by waterfowl, which feed heavily on *Vallisneria*.

Other plants have evolved similar reproductive strategies that incorporate some of the aspects of *Vallisnerias* reproductive behavior described above. In *Hydrilla*, for example, the male flowers abscise and float free on the surface and in *Nymphaea* and numerous other plants, the peduncles coil to withdraw the developing fruits beneath the water (Fig. 9.1).

In anemophilous and entomophilous species, numerous devices have evolved for keeping the flower above the water. In the water lilies, the flower is waxy and bowl-shaped and floats by itself in the manner of a small boat. In other species such as the water milfoils, the apical end of the stem becomes rigid and protrudes above the water and small flowers are borne on it (Fig. 9.2). The submersed, lax portions of the stems form large tangled mats that lie just beneath the surface and provide a stable platform to support the emersed tips and prevent them from being blown over by the wind.

Some of the bladderworts of the genus *Utricularia* provide a stable platform in a different manner. The last whorl of leaves at the end of the

Fig. 9.1.   Coiled peduncles on pollinated flowers of fragrant water lily. The coiling of the peduncle retracts the developing fruit and seeds beneath the surface where they are presumably less likely to be eaten.

Fig. 9.2.   A dense stand of water milfoil showing lax underwater stems with
stiff apical tips protruding above the surface. The stiff tips hold tiny axillary
flowers above the water where pollination occurs.

stem becomes enormously enlarged and full of spongy tissue, full of air
pockets, known as aerenchyma. These leaves float on the surface and
radiate out from the stem like the spokes on a wheel (Fig. 9.3). The scape
with the flower at the end arises from the center of these floating leaves
and is supported and stabilized by them. Fanwort has also evolved a
system for using leaves to help maintain the flowers above the water.
Typical fanwort leaves are flat, fanshaped, finely divided, and submerged. At the time of flowering, a few leaves grow at the end of the stem
that are completely different. They are very small, peltate, and undivid-
ed. They float on the surface and, as in the previous examples, provide
support and stability to the flower and flower stalk, which are above the
water.

Coontail provides a good example of hydrophilous pollination. The
flowers are small and are borne underwater in the leaf axils with male and
female flowers at separate nodes. At maturity the individual anthers of
the male flowers abscise, float free from the rest of the flower, and rise to
the surface where the pollen is shed. Individual pollen grains sink slowly
through the water and some, by chance, land on the stigmas of female
flowers which remain attached to the parent plant.

Fig. 9.3. A view straight down on a blooming blad-derwort plant, showing the inflated whorl of leaves which serves as a stable, floating platform for the emergent flower stalk.

Vegetative Reproduction

In general, vegetative reproduction is more common and more impor-tant to survival among aquatic plants than it is among terrestrial plants. Some species rarely produce viable seeds, but even those that do, fre-quently exhibit a great deal of reliance on vegetative reproduction, and the seeds serve more as a back-up reproductive system to insure survival in the event of a disaster.

For example, Patten (1954) studied Eurasian water milfoil in New Jersey and found that it produced large numbers of viable seeds but that they never developed into new plants in nature. He found that all new plants developed vegetatively, primarily by stem fragmentation. Coontail grows profusely in England but does not produce any viable seeds (Guppy 1894). Coontail propagates almost entirely in that country by means of axillary buds that develop in the spring from fragments of previous year's stems that otherwise appear dead. Fanwort produces viable seeds within its original range in the southeastern United States but not in the northeastern United States, where it is adventive (Riemer and Ilnicki 1968).

LEAVES

Leaves on aquatic plants may be of three basic types: aerial, floating, and submersed. It is not uncommon to find two of these types occurring on the same plant at the same time.

Aerial Leaves

Aerial leaves of emergent and free-floating species are morphologically and anatomically very similar to leaves of terrestrial plants. This is not surprising, considering their common ancestry and their common aerial environment. Although they are normally surrounded by air of high relative humidity, aerial leaves of aquatic plants may transpire at a substantial rate. As in terrestrial plants, the bottom surface of the leaf contains stomata and the epidermis is covered with a waxy cuticle.

One major difference between emergent aquatic species and terrestrial species is that the emergent species are submerged for a short time during their early development, after they germinate or begin vegetative growth in the spring, and before they reach the surface. At this time they may be subjected to very low oxygen levels or even to anaerobic conditions. During this initial stage of growth, some plants have the capability of respiring anaerobically. Young cattail leaves and seedlings of arrow arum are representative of this type of plant. They can respire anaerobically for several days, which presumably is sufficient time to allow them to reach the surface or at least to get further away from the substrate where oxygen levels are usually lowest. After reaching the surface, these plants require oxygen, just as any terrestrial plant does.

Floating Leaves

Floating leaves, as defined in this chapter, are those whose blades float in a horizontal position with their upper surfaces exposed to the air and their lower surfaces exposed to the water. The term does not include the leaves of floating rosettes whose petioles may be in the water but whose blades extend into the air, such as those of water hyacinth and water lettuce.

Since floating leaves are exposed to two markedly different environments at the same time, they require some special adaptations. Parallel evolution has resulted in very similar physical characteristics in most floating-leaved species; even those as distantly related as dicots and monocots. Some of these physical adaptations to the floating habit of growth are discussed below.

Leaf Shape. Almost all floating leaf blades have evolved toward a circular shape with an entire margin. Some species are closer to this ideal than others, but the trend is usually quite evident. This shape apparently provides maximum protection against tearing when leaves are subjected to wind and wave action. Leaves with deep sinuses along their margins would provide points of concentration for stresses and strains and tears would be more likely to develop than they would in circular leaves without sinuses or other irregularities.

Leaf Texture. Floating leaf blades invariably have a tough, leathery texture that affords additional protection against tearing and perforation. Since they are supported from beneath by the water, these leaf blades are not able to "give" or bend under the impact of heavy rain or hail and a thin, papery texture would allow large raindrops or hailstones to penetrate.

The upper surface is usually covered with a heavy, waxy cuticle that sheds water and thus helps to prevent immersion. The lower surface of the leaf, which is in contact with the water, is usually devoid of a waxy cuticle or has only a very thin layer. A few species have a covering of dense hairs that trap air among them when submerged, thus preventing wetting of the leaf surface.

Other Adaptations of Leaf Blades. Among other adaptations are the placement of stomata and the presence of internal air spaces. Unlike the leaves of terrestrial plants, floating leaves have their stomata entirely on the upper surface where they communicate with the atmosphere rather than the water. If any stomata occur on the bottom surface of the leaf, which is rare, they are nonfunctional.

Most floating leaves have exceptionally large air spaces in the mesophyll beneath the palisade layer. This presumably provides flotation. They also have special air spaces called lacunae, which will be discussed in greater detail in the section on stems, later in this chapter.

Petioles. The petioles of rooted plants with floating leaf blades exhibit considerable modification. They are usually long, thin, and very flexible. In almost all cases the petiole is longer than the minimum length required to allow the blade to just reach the surface. This additional length allows the leaf blade to float up and down with fluctuating water levels so that it will not be innundated by a sudden rise. The additional length also allows the leaf blades to spread out on the surface so that they do not overlap and shade each other. Leaves of terrestrial plants have three dimensions in which to arrange themselves in order to intercept a maximum amount of sunlight. Floating leaves have only two dimensions in which to accomplish the same result. They must, therefore, be able to spread apart on the surface, even when the petioles arise from essentially a single point as in the water lilies.

If floating leaf blades become submerged, as in the case of a flood, petioles immediately begin growing at a rapid rate until the blades once again reach the surface, at which time growth ceases. The physiological cause of the initiation of growth and its subsequent cessation are unknown but could involve light intensity or quality, oxygen and/or carbon dioxide concentrations, or the ability of the leaf to transpire (Sculthorpe 1967).

Submersed Leaves

In many respects, submersed leaves are the antithesis of floating leaves. They differ from each other in many respects even though both types may occur on the same plant at the same time as in the case of some pondweeds of the genus *Potamogeton*. Some of the more important characteristics are discussed below.

Leaf Shape.   Most submersed leaves have a shape that gives them a very high surface-to-volume ratio. This large surface area permits greater exchange of gases between the plant tissue and the water. This is important to plants that are not in direct contact with the atmosphere and must obtain their oxygen and carbon dioxide from the supply dissolved in the water. Leaves of most submersed plants fall into one of two broad categories, based on their shape.

One category consists of long, extremely thin leaves with entire margins. Leaves of this type are often so thin as to be transluscent and so long and narrow that they are described as being "tape-like," "ribbon-like," or "thread-like" (wild celery, Fig. 7.32). Other leaves in this general category are not quite as long and are frequently described as "strap-shaped" or "tongue-shaped" (elodea, Fig. 7.31).

The other general category consists of leaves which are deeply dissected along the margins or compound leaves in which the blades are separated into individual segments called leaflets. Leaflets may be flattened as in fanwort (Fig. 7.23) or hairlike as in water milfoil (Fig. 7.30). This finely divided condition exposes a large surface area to the environment while reducing resistance to the current.

Leaf Texture.   Underwater leaves tend to be thin and very pliable. Tissues for mechanical support such as collenchyma and sclerenchyma are drastically reduced or completely absent and even the vascular elements are lacking in lignin. Support is provided instead by the water that surrounds the leaf and by aerenchyma within the blade, which makes the leaf buoyant and prevents it from hanging limply in a vertical position. The lack of rigidity in submersed leaves doubtlessly helps to prevent mechanical damage in swift currents.

The surface texture of submersed leaves is always smooth and without pubescence. A waxy cuticle is absent, which facilitates the diffusion of dissolved gases and solids into and out of the tissue. Water loss through the leaf surface is obviously not a problem. The absence of a cuticle is the logical extension of a phenomenon that has been observed in terrestrial plants, in which the exposure to high humidity results in a reduction in the cuticle.

Heterophylly

Heterophylly is a complex phenomenon that may be simply defined as the presence of more than one kind of leaf on the same plant. While heterophylly is not unknown among terrestrial plants, it is far more common and the differences among the different leaf types tend to be greater in aquatic plants. Heterophylly is most frequently manifested in the form of floating leaves and submersed leaves on the same plant (Fig. 9.4), or as submersed leaves and aerial leaves on the same plant (Fig. 9.5). These different leaf types, as can be seen from the descriptions above, are distinctly different in form and texture.

Other variations in leaf form may occur as a result of differences in water depth, rate of current flow, and possibly other environmental variables. An extreme case is the production of "land forms" by many species. If water milfoil, for example, is left stranded by receding water levels, the typical submersed shoots will dessicate and die but new shoots will be produced that have an entirely different habit of growth. The new shoots are called land forms and only grow to a height of about 2.5 cm (1 in.). The stems are relatively rigid and the internodes are very short when compared to the typical submersed form. Leaves are smaller with fewer segments but the segments are broader and thicker than in the water form (Arber 1963). The land form of *Myriophyllum humile* is pictured in Fig. 9.6.

STEMS

Stems of submersed aquatic plants exhibit many of the same adaptations to their underwater existence as do the leaves . Among these are a lack of rigidity, little or no cuticle, and the presence of aerenchymous tissue. Most submersed stems contain chlorophyll, even in their epider-

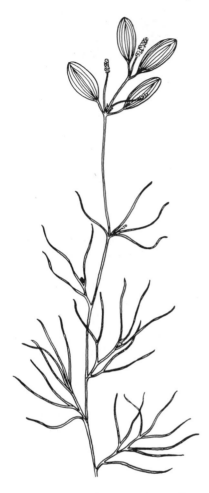

Fig. 9.4. Heterophylly in *Potamogeton diversifolius*. This species has flattened, elliptic, petioled, floating leaves and linear, sessile, submersed leaves. *Courtesy of N.J Coop. Extension Service.* (0.66×).

mal cells, and physiologically they function much like a leaf. The role of the stem in physical support is greatly reduced because of the buoyancy of the plant.

The vascular system is typically reduced in submersed stems (and leaves as well), but there is considerable evidence that a net upward flow, similar to a transpirational stream, does occur, despite the lack of normal transpiration at the leaf surface. This has been demonstrated using dyes and radioactive tracers.

ROOTS

Plant roots serve two basic functions: anchorage and the absorption of water and dissolved substances. Both functions occur in hydrophytes but

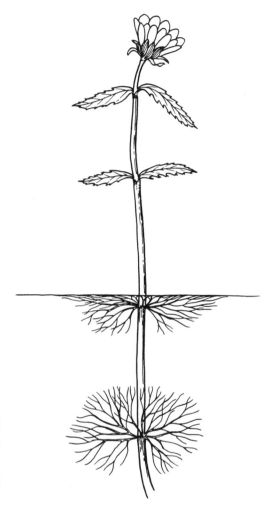

Fig. 9.5. Heterophylly in *Bidens beckii*. This species has submersed, finely divided, compound leaves and emersed, simple, toothed leaves. *Courtesy of N.J. Coop. Extension Service.* (0.66×).

some modifications are evident. Although the root system is reduced in many species and entirely lacking in a few, it is well-developed and highly functional in others. There is so much variation that it is difficult to draw any general conclusions about the root systems of aquatic plants.

Anchorage

The need for anchorage varies with the habit of growth and the habitat. Emergent species with most of the plant body in the air require just as much anchorage as terrestrial plants. Submersed species growing in swiftly flowing water require anchorage in order to prevent being swept away by the current, but submersed species growing in still waters

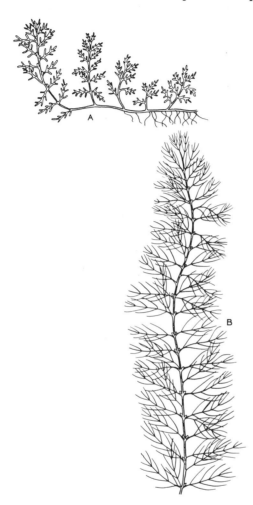

Fig. 9.6. Two distinct growth forms in *Myriophyllum humile*. (A) The land form which occurs when submerged plants are stranded by a receding waterline. (B) The typical submerged form. *Courtesy of N.J. Coop. Extension Service.* (0.5×).

require minimal anchorage. Free-floating species are not anchored by roots at all, although they may form solid interconnected mats that are anchored in place by catching on stumps, overhanging trees, or other obstructions. In this manner, plants such as water hyacinth can persist and thrive in slow-moving rivers and canals. Free-floating plants such as the duckweeds, which do not form strongly interconnected mats, are restricted to nonflowing waters or to narrow areas along the margins of streams where the current is negligible and there is sufficient debris to catch and hold the plants.

Absorption

The role of roots in the uptake of ions doubtlessly varies among different kinds of aquatic plants. Emergent species absorb nutrient ions and water through their roots in the conventional manner. The site of uptake by submersed plants, however, has been in question for many years, and it has only been recently that some definitive answers have been forthcoming. While ion absorption is possible through leaf and stem surfaces as well as root surfaces, the relative amount of uptake by each organ is poorly understood. It is a question of vital importance since uptake by stems and leaves reduces the concentration of that particular ion in the free water, while uptake by the roots reduces the concentration in the sediments. In the former case, aquatic plants serve as sinks and in the latter case they serve as pumps, moving elements from the sediments into the free water by excretion or on decomposition of the plants.

Carigan and Kalff (1980) studied the uptake of phosphorus by eight common species of aquatic plants and concluded that they depend overwhelmingly on the sediments for their supply of this nutrient and that all submersed aquatic species should be looked on as potential phosphorus pumps. The relative importance of root uptake versus shoot uptake for other nutrients is not understood as well and a great deal more work needs to be done on this problem.

LACUNAE

The lacunar system is an adaptation to underwater environments that is found in a wide variety of plants, even those only distantly related to one another. It consists of an interconnecting system of gas-filled canals and spaces that permeate virtually all submersed stems, petioles, and leaf blades. The individual lacunae do not have walls of their own and are not "structures" in the sense that the vascular elements are. Rather, they are open spaces that develop between other tissues (Fig. 9.7). They begin as

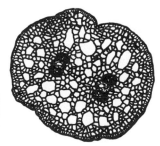

Fig. 9.7. Cross-section of a stem of fanwort showing vascular elements (dark areas) and lacunae (large open spaces). (Tracing of a photomicrograph by author).

tiny cracks between cells, very close to the growing tips of submersed organs. The tiny cracks enlarge with time, perhaps due in part to the pressure of the gas within them. In many plants, open, uninterrupted lacunar spaces extend from the tips of each leaf, through the petioles and stems, and down into the buried rhizomes and roots. They may form a very appreciable percentage of the volume of some organs. For example, 60% of the rhizome of *Cladium mariscus* may consist of lacunar space (Conway 1937, 1942, as reported by Hutchinson 1975).

Lacunae serve as reservoirs for gases and, in some cases at least, as a transportation system for gases. It has long been known that gases actively flow through lacunae and are not restricted to movement by diffusion alone. Rutner (1953) pointed out that a floating water lily leaf will take in air from the atmosphere and force it out of the end of a severed petiole beneath the surface. He also pointed out that the bubbling stops when the leaf blade is shaded. More recently this phenomenon has been studied in depth by Dacey (1980) in the yellow water lily, *Nuphar luteum*. He has shown that during periods of sunshine this plant draws air into the younger leaves against a pressure gradient. It then moves down through the lacunae of the petioles into the rhizome and eventually back up through the older leaves into the atmosphere. This ventilation system brings oxygen to the rhizome buried in the anaerobic sediments and carries carbon dioxide from the rhizome to the atmosphere. The energy for pumping air against a pressure gradient is derived from the heat of the sun, and not from photosynthesis.

It is not known how widespread this ventilation phenomenon is among aquatic plants, but it is obviously possible only among those species having both aerial and submersed parts. Lacunae are present in virtually all submersed species but their primary function appears to be gas storage.

BIBLIOGRAPHY

ARBER, A.   1963.   Water Plants. Hafner Pub. Co., New York.

CARIGAN, R., and KALFF, J.   1980.   Phosphorus sources for aquatic weeds: water or sediments? Science **27**, 987−989.

DACEY, J.W.H.   1980.   Internal winds in waterlilies: an adaptation for life in anaerobic sediments. Science **210**, 1017−1019.

DE WIT, H.C.D.   1978.   Pollination problems in aquatic plants. Acta Bot. Neerlandia **27**, 91.

GUPPY, H.B.   1894.   Water plants and their ways. Sci. Gossip. **1**, 145−147; 178−180; 195−199.

HUTCHINSON, G.E.   1975.   A Treatise on Limnology. Vol. III, Limnological Botany. John Wiley & Sons, New York.

PATTEN, B.C. JR. 1954. *Myriophyllum spicatum* L in Lake Musconetcong, New Jersey: its ecology and biology with a view toward control. M.S. Thesis. Rutgers University., New Brunswick, New Jersey.

PRESCOTT, G.W. 1980. How to Know the Aquatic Plants, 2nd Ed. Wm. C. Brown Co., Dubuque, IA.

RIEMER, D.N., and ILNICKI, R.D. 1968. Reproduction and overwintering of cabomba in New Jersey. Weed Sci. **16**, 101–102.

RUTNER, F. 1953. Fundamentals of Limnology, D. G. Frey and F.E.J. Fry (Translators). Univ. of Toronto Press, Toronto, Ontario, Canada.

SCULTHORPE, C.D. 1967. The Biology of Aquatic Plants. St. Martins Press, New York.

Part III

Aquatic Plants
in Relation to Humans

10

Mankind's Interests in Manipulating Aquatic Plant Populations

Human beings are rarely satisfied with their natural surroundings as they find them. They continually attempt to modify them for their own benefit by increasing or maximizing what appears useful or beneficial and decreasing what appears useless or harmful. Aquatic plants are not excepted from this human behavior. Part III of this book will deal with human activities intentionally and specifically undertaken in order to modify aquatic plant populations in some way. Human activities which unintentionally alter aquatic plant populations are discussed in Part I of this book.

Activities intended to alter aquatic plant populations can be categorized as those designed to benefit and increase desirable species and those designed to control or eliminate undesirable species. In the United States (and probably worldwide as well) far more time, effort, and money are devoted to the control of undesirable species than to the introduction, cultivation, or encouragement of desirable species. On rare occasions, a management program has been specifically designed to do both.

REASONS FOR INCREASING CERTAIN SPECIES

Wildlife

One of the major reasons for wanting to establish or increase certain species of aquatic plants is to attract or benefit wildlife— especially waterfowl. A major facet of waterfowl management is habitat maintenance and habitat improvement, both of which involve management of aquatic plant populations. Different aquatic plant species vary tremendously in their value to and utilization by waterfowl. Some species are highly desirable while others are useless and still others can be harmful to waterfowl and waterfowl management programs. Eurasian water milfoil, for example, is not only of little value in itself, it is an aggressive and competitive species that crowds and sometimes completely eliminates valuable species such as widgeongrass, wild celery, and various species of pondweeds (Steenis and Stotts 1961).

It is imperative that waterfowl managers be able to identify aquatic plants and that they have a sound working knowledge of the plants' requirements, life histories, and population dynamics. It is interesting to note that much of our early knowledge of the ecological requirements of aquatic plants came from the observations and field experiments of water- fowl biologists, not botanists.

Other species of wildlife such as moose, muskrats, rails, and songbirds also use aquatic plants for food, cover, nesting materials, and nesting sites (Martin *et al.* 1961), but the management of these animals has never become so dependent on aquatic plant manipulation as has the manage- ment of waterfowl. Fisheries biologists rarely, if ever, introduce new plant species into a body of water or undertake activities to benefit existing vascular plant species as parts of their programs for management of fish populations. They do, however, sometimes fertilize small bodies of water with ordinary agricultural fertilizer in order to increase algae populations, which increases the overall productivity of the system and ultimately increases fish production. We have seen in Chapter 8 that flowering plants, ferns, and mosses have little impact on the immediate food relationships in a pond or a lake. For this reason, fisheries biologists are much more likely to consider vascular aquatic plants as pests rather than something to be nurtured and encouraged.

Esthetics

Another reason for attempting to increase certain species of aquatic plants is for their esthetic and ornamental value. Many floating and emergent hydrophytes have attractive flowers and foliage, which make them useful for decorating pools and ponds in parks, golf courses, home grounds, and other sites. They are also occasionally used in indoor pools in shopping malls, restaurants, and similar public places. Water lilies are commonly used for such purposes as are flowering emergent species such as pickerelweed and arrowhead. The free-floating water hyacinth, which has gained notoriety as one of the worst weeds in the world, was originally brought into this country as an ornamental plant for use in a pool at a Cotton States Exposition in New Orleans in 1884 (Tabita and Woods 1962). Papyrus and related plants of the genus *Cyperus* are also used as ornamentals because of their delicate, graceful, emergent foliage.

A small but active industry exists in the United States today that specializes in raising and selling aquatic plants for decorative purposes. In many cases, the same businesses provide plants for the aquarium trade, although many of the submerged plants sold for use in aquariums are collected from wild or semi-wild populations.

Some people do not introduce ornamental plants into their ponds, but appreciate and enjoy the more attractive of those native species that appear by natural colonization. These people, too, are anxious to protect and perhaps to increase the number of such plants.

A final reason for wishing to increase populations of aquatic plants is that some species are used for human consumption. This topic will be discussed in greater detail in Chapter 14.

REASONS FOR CONTROLLING CERTAIN SPECIES

If we are going to discuss the control of aquatic weeds, we should pause for a moment and decide which plants are included in the term "aquatic weed." The most frequently quoted definition of the word weed is "a plant out of place." Many other definitions have been used such as "a plant whose potential for harm is greater than its potential for good," "a useless or troublesome plant," and "a plant growing where it is not wanted." These definitions and all other valid definitions of the word weed have one thing in common—they all show a relationship between plants and people. If there were no people on the planet earth, there would be no weeds. Weeds are weeds only because in some way they are causing a problem or pose a potential problem to someone. The problem may be of major proportions or a trifling nuisance, but if there is no problem or potential problem, then there is no weed.

No plant is a weed merely because it belongs to a particular genus and species. Similarly, no plant can escape the appellation "weed" merely because it belongs to a genus and species that is normally thought of as being a desirable plant. No one can deny, for example, that a corn plant is a weed if it is growing on a green of a golf course. Sago pondweed is very desirable in a duck marsh, but it is a weed when it blocks the flow of water in an irrigation ditch.

Even within a single body of water, a particular stand of plants may be weeds to one person and desirable plants to someone else. A patch of lily pads may be good bass habitat in the eyes of a fisherman and an unwanted nuisance in the eyes of a swimmer. The word weed, therefore, may mean different things to different people. This occasionally causes problems in the planning of aquatic weed control programs, because some people want all of the plants eliminated, other people want certain species removed selectively, and still other people do not want any of the plants removed. In situations of this nature, a compromise must be reached before a weed control plan can be formulated and implemented.

The reasons for wishing to control aquatic weeds are numerous and varied. The relative importance of these reasons differs in different parts of the country and in different parts of the world. Some of the more important of these reasons will be discussed in the following paragraphs.

Recreation

Aquatic weeds interfere with boating, water skiing, swimming, fishing, and just about every other recreational use that mankind makes of water. Dense stands of aquatic vegetation can render a lake or pond totally useless for most recreational purposes (Fig. 10.1). In heavily populated regions of this country, such as the Middle Atlantic States, the restoration or preservation of the recreational value of bodies of water is the single most important reason for undertaking aquatic weed control operations. In other parts of the country and in many parts of the world

Fig. 10.1. A mass of hydrilla on an oar. Dense growths of weeds such as this can make boating and other recreational activities difficult or impossible.

outside of the United States, aquatic weed programs are rarely undertaken to enhance recreational potential.

Esthetics

A different but closely related reason for controlling aquatic weeds is that they impair or completely destroy the esthetic value of all types of waters. A small pond built for landscaping purposes can change from an asset to a detriment by growths of unsightly weeds. Similarly, a picture window in a private home or the dining room of a country club that overlooks a weed-choked lake whose shoreline is littered with piles of decaying vegetation blown ashore by the wind is not attractive. Some people have actually sold their homes and others have petitioned for property tax relief because of unsightly, foul-smelling aquatic weed problems in lakes or bays on whose shores their houses were situated. It is difficult to put a monetary value on the esthetic quality of a lake, pond, or stream, but to people such as those described above and to park managers and golf course superintendents, esthetics are extremely important.

Mosquito Control

A more practical reason for controlling aquatic weeds is the fact that they increase mosquito populations in areas surrounding the waters in which the weeds occur. Aquatic plants provide the stagnant water and quiet surface conditions that mosquito larvae require. In addition, aquatic plants protect mosquito larvae from predation by fish and other predators by providing hiding places for the larvae and by interfering with the movement of predators. With reduced predation there is an increase in the number of larvae that survive to become adult mosquitos. Furthermore, some mosquitos are absolutely dependent on aquatic plants for their survival. The larvae of these genera do not obtain their oxygen from the atmosphere at the water's surface as do the larvae of most mosquitos. Instead, they pierce the stems of aquatic plants and obtain their oxygen from the lacunae within (Mulrennan 1962). Numerous studies have shown that reducing aquatic plant populations will reduce mosquito populations in the surrounding area.

Mosquitos are not only a nuisance but a major public health problem in tropical and subtropical parts of the world. Many of the world's most extensive and dreaded diseases are transmitted to humans by mosquito vectors. Among the more serious of these diseases are malaria, yellow fever, encephalitis, dengue, and filariasis. Public health programs designed to combat these diseases frequently include aquatic weed control operations.

Snail Control

Snails are also involved in perpetuating the life cycles of organisms that cause human diseases. Many of these snail-related diseases are less familiar to us than the mosquito-borne diseases, but they are, nonetheless, among the most serious and widespread afflications of mankind. One of them, schistosomiasis, afflicts an estimated 240–300 million people and causes 2–4 million deaths each year (Cardarelli 1976). This makes it one of the most important human diseases in the world. Other snail-related diseases important on a worldwide basis are fascioliasis, angiostrongyliasis, clonorchiasis, paragonimiasis, and fasciolopsiasis.

The suppression of snails in order to reduce the incidence of these diseases is yet another reason for controlling aquatic vegetation. Aquatic plants provide food for snails as well as supplying hiding places to protect them from predation. Some species of snails use emergent plant stems as places to deposit their eggs while other species of snails deposit them on submersed plant surfaces. Aquatic weed control is an important part of the fight against disease-carrying snails in the tropics.

Fisheries Management

Dense stands of aquatic plants can be troublesome to fisheries biologists in two distinctly different ways. First, they can interfere with the harvest of fish at hatcheries and fish farms. This is a purely mechanical problem of the weeds interfering with the movements of nets and other harvesting gear. If ponds are drained during the harvesting operation, fish may become trapped in masses of weeds and never get down to the "collection basin," which is the deepest part of the pond into which fish are concentrated during harvest. Fish that are trapped in weeds and left stranded on the exposed parts of the pond bottom, die and are never harvested.

The second major problem that aquatic weeds pose to fisheries biologists is one that was discussed in Chapter 8. It is the stunting of fish populations caused by the lack of adequate predation on young fish and the over-population that results. The fish in these stunted populations are not of harvestable size and will never grow larger unless their population densities are reduced. Fisheries managers, therefore, are interested in controlling excess vegetation in order to increase the amount of predation on young fish.

On the other hand, they rarely wish to eliminate all vegetation because over-predation is also a possibility. In addition, plants provide shade, which is utilized by some game fish such as large mouth bass, and they provide spawning areas for others, such as eastern chain pickerel. The complete absence of vascular plants may be desirable in some hatchery situations, but not in a lake that supports a warmwater sport fishery.

Navigation

Certain types of aquatic plants, when growing in extremely dense stands, can interfere with or even prevent navigation, and this constitutes still another reason for controlling their populations. In this country, the United States Army Corps of Engineers has the responsibility for maintaining navigable waters, and they get deeply involved in aquatic weed control in the course of fulfilling this responsibility. One of the most outstanding examples of aquatic weeds interfering with navigation is the occurrence of water hyacinth in Florida. Although this plant grows in the form of free-floating rosettes, they crowd together into vast floating mats that extend from shore to shore and quickly render a river impassable to boat traffic. Were it not for the constant maintenance activities of the Corps of Engineers, many of the rivers in Florida and other southeastern states would be completely clogged and useless for commerce or recreational boating.

Irrigation

Farmers, too, have a vital interest in suppressing hydrophytes because they create problems with irrigation systems. In small systems where farmers pump water from privately owned ponds or reservoirs, weeds get drawn into the intakes, thereby reducing or cutting off the flow of water. Some people have built elaborate screening devices to keep weeds away from intakes but they are only partially successful at best. If the mesh of the screen is too small, the screen soon plugs up and water flow is once again restricted. If the mesh is too large, fragments of weeds pass through and reach the intake anyway.

In large-scale irrigation systems such as those used in the western United States, water is delivered to the growers through extensive systems of canals operated by public or private organizations. The presence of weeds in these canals slows down the rate of flow to the point where sufficient water can no longer be delivered to the users. Sago pondweed is one of the major problem species in these canals, although other species also cause problems. The federal government gets involved in aquatic weed control in some of these canals because the Bureau of Land Management is charged with maintaining approximately 173,000 miles of these canals in 17 western states. An estimated 63% of this mileage is seriously infested with aquatic weeds, and in 1975 the Bureau of Land Management spent over $10 million for aquatic weed control, which represented more than 30% of the total cost of operation and maintenance (U.S. Dept. of Agric. 1976).

In addition to these widespread problems, aquatic weeds also cause numerous problems more local in their impact but just as serious to those people who happen to be affected. A good example of this type of local or regional problem is the sudden and dramatic infestation of Eurasian water milfoil that occurred in Chesapeake Bay in the the late 1950s and lasted throughout the 1960s. This infestation created many of the problems just described, but also created a rather unique problem of tremendous economic impact to those concerned. By slowing down and diverting the currents that brought the necessary oxygen and food to oysters, the milfoil smothered thousands of acres of oyster beds and took them completely out of production. In 1964, for example, over 200,000 acres of Chesapeake Bay were infested with Eurasian water milfoil. Of this total, approximately 1600 acres consisted of charted natural oyster beds. The loss in revenue for that year was estimated to be $1,280,000 in Maryland waters alone (Steenis and King 1964). In addition, there was a smaller loss in that portion of the bay lying within Virginia's boundaries. Additional losses were incurred in the clamming and crabbing industries.

These examples should make it clear that there are numerous and varied problems associated with aquatic plants, requiring some kind of remedial action to relieve the problem. The next three chapters will discuss in some detail the methods of control that have been utilized over the years and the advantages and disadvantages of each. Before going on to those chapters, however, we should have a clear understanding of what is meant by "control" in the phrase "aquatic weed control".

CONTROL, ERADICATION, AND PREVENTION

Aquatic weed control may be defined as the reduction of the population density, the vigor, or both, of an aquatic weed population, to an acceptable level. Let us analyze this definition by looking at it one piece at a time. First of all, if there is going to be aquatic weed control, there has to be a population of aquatic weeds. By our earlier definition of aquatic weed, there must also be a problem or a potential problem. "Control" means that we are going to reduce the number of weeds (their population density), or else we are going to reduce the vigor of the individual plants, or both, to the point where the problem no longer exists or at least is reduced to the point where it is acceptable.

"Eradication," on the other hand, is the total elimination of the weed from a particular body of water, a particular state, or some other geographic area. Eradication is rarely possible or economically feasible except in extremely small areas. In many cases it is not even practical to eradicate a weed species in a farm pond that is only an acre or two in surface area.

When eradication can be achieved, it is certainly an example of control but control by no means implies eradication. This distinction should be kept in mind while reading the remaining chapters in this book.

Prevention is, in a sense, also a form of control. It is a form of control used before the problem occurs. Human nature being what it is, few people worry about weed problems they do not have yet, so prevention of aquatic weeds gets very little attention. Prevention can best be achieved by not introducing any new plant into a body of water, intentionally or unintentionally, unless one knows absolutely what it is and what its potential is for becoming a weed problem. Countless numbers of people have purposely planted new species of plants in lakes, ponds, and even rivers, only to regret it later. Other people unwittingly introduce new species to a body of water when they transport their boats overland on boat trailers from one lake or river to another. Plants get caught on the boat or the trailer and are carried to the next lake or river where they are washed off to establish themselves in the new location. This has been documented in many areas, and, on occasions, when a major effort was under way to prevent the spread of a particular plant species, people were hired to inspect all boats being launched at certain launching sites to insure that they were free of the weed in question.

Another preventative measure can be taken at the time a new pond is being constructed. Care should be taken to insure that the sides of the pond slope down steeply from the water's edge so that there will be a miniumum area of shallow water around the perimeter of the pond. This shallow, marginal area is frequently the first area to be colonized by aquatic weeds, and by keeping it as small as possible, future problems may be somewhat less severe in magnitude. Emergent species such as reeds, rushes, sedges, cattails, and aquatic grasses become particularly troublesome in these shallow, marshy areas, which are too wet to mow and too shallow to get a boat into for effective spreading of herbicides.

BIBLIOGRAPHY

CARDARELLI, N. 1976. Controlled Release Pesticide Formulations. CRC Press, Cleveland, OH.

MARTIN, A.C., ZIM, H.S., and NELSON, A.L. 1961. American Wildlife and Plants, A Guide to Wildlife Food Habits. Dover Publications, Inc., New York.

MULRENNAN, J.A. 1962. The relationship of mosquito breeding to aquatic plant production. Hyacinth Control J. 1, 6–7.

STEENIS, J.H., and KING, G.M. 1964. Report of Interagency Workshop Meeting on Eurasian Water Milfoil, Annapolis, MD, Feb. 20, 1964.

STEENIS J.H., and STOTTS, V.D. 1961. Progress report on control of Eurasian water milfoil in Chesapeake Bay. Proc. Northeastern Weed Control Conf. **15**, 566–570.

TABITA, A., and WOODS, J.W.   1962.   History of hyacinth control in Florida. Hyacinth Control J. **1**, 19–22.

U.S. DEPT. OF AGRICULTURE.   1976.   Weed Control Technology for Protecting Crops, Grazing Lands, Aquatic Sites, and Noncropland. Agric. Research Service, National Research Program No. 20280. Washington, DC.

11

Biological Control of Aquatic Vegetation

Biological control has been defined as "the action of parasites, predators, or pathogens in maintaining another organism's population density at a lower average level than would occur in their absence." A simpler definition is "the use of one type of living organism to control another." We shall refer to the organism doing the controlling as the control agent. For the sake of convenience, we will refer to the organism being controlled as the pest, but be aware that organisms other than pests are also subject to biological control and that this happens all the time.

As the first definition implies, the control agent may occur in many forms. It may be a fungus, bacterium, or virus that causes a disease in the pest. The disease may be fatal, it may prevent reproduction by the pest, or it may weaken the pest to the point where it succumbs to other factors. The control agent may also be an animal that feeds on the pest. It may devour the pest in its entirety as in the case of a bird eating an insect, or it may eat slowly as in the case of beetles feeding on the leaves of a plant. The control agent may not affect the pest directly at all; it may share the pest's environment and compete with the pest for nutrients, light, water, space, or other necessities of life. If the control agent is a strong enough competitor, the pest will be suppressed or even eradicated.

Biological control is not something that mankind invented. Natural biological control is going on constantly in all ecosystems, and the organisms being controlled are not necessarily those organisms we call pests. In fact, if it were not for natural biological control, the majority of all plant and animal species on earth would become pests because of the enormous populations that would ensue. In the remainder of this book, however, we are going to concern ourselves only with biological control mediated by mankind. The term biological control from here on will mean biological control that is purposely undertaken, usually by introducing the control agent into a new geographic area where the pest is causing a problem and where that particular control agent does not presently live.

REQUIREMENTS OF CONTROL AGENTS

Biological control is a very attractive pest control strategy, in theory, for a number of reasons. It is economical in the long run because once the control agent has been permanently established in the new area, nothing further needs to be done. The control agent will remain in the new area, constantly suppressing the pest population. There are no additional labor costs, no additional equipment costs, and no additional supplies to be purchased. Furthermore, there is no danger of chemical residues in the environment that might pose a threat to non-target organisms, including humans.

In actual practice, this Utopian situation is rarely attainable. Several serious drawbacks exist. For one thing, a control agent must be found that has all of the following attributes:

1. It must attack only the target plant or plants and not desirable plants or animals.

2. It must be able to survive in the new environment into which it is to be introduced.

3. It must be capable of reducing the pest-created problem to economically or esthetically acceptable levels.

A control agent meeting all three of these requirements may not exist for every pest or even for every aquatic weed. In fact, it would be quite illogical to assume that for every aquatic weed, there is a control agent somewhere in the world.

Let us look at each of these requirements in greater detail, specifically as they apply to aquatic weeds. First, there are relatively few organisms that feed exclusively on a single species of plant. There are probably more pathogens that are host-specific, but still, great care must be taken to insure that an organism will not be introduced into an area where it will become a greater pest than the the one it was supposed to control. One must be constantly aware of the population explosions that can occur when any organism is moved into a new environment.

The second requirement is that the control agent must be able to survive in the new environment into which it will be placed. This means that it must be compatible with the new climate, the new physical environment, and the new biological environment. In particular, there must not be any organisms in the new environment that will suppress the control agent's population to the point where it cannot function as intended. In other words, there cannot be biological control of the control agent.

The final requirement is that the control agent must be capable of reducing the aquatic weed population to an acceptable level. Merely feeding upon the weed or causing it to become diseased is not enough. It must reduce the problem to an acceptable level, or, at the very least, alleviate the problem to some degree. This is probably the single most overlooked requirement among laymen who advocate the use of biological control without a full understanding of all of its ramifications. No matter how much feeding an organism does on a pest, it is not a control agent if it does not achieve control. Merely attacking the pest is not sufficient. A good example of an organism that attacks a pest but does not control it is *Galerucella nymphaea*, a chrysomelid beetle that feeds on spatterdock, a yellow-flowered water lily with leaves frequently elevated in the air rather than floating on the surface of the water.

This beetle is abundant in New Jersey, where it feeds very heavily on those portions of spatterdock plants above the waterline. The beetle's entire life cycle is intimately associated with the plant. Both the adults and the larvae feed on it and the eggs are laid on it as well. By the middle of August, many stands of spatterdock are so badly eaten that it is actually impossible to find a whole leaf. The majority of leaf blades are more than half missing and the remaining portions are full of holes. The margins of the holes and of the blades themselves are brown, necrotic, and curled. Many leaf blades have been eaten away entirely, leaving a badly chewed stump of a petiole protruding from the water. It appears to the casual observer that the stand is almost dead and will soon be gone. The fact is, however, that the plants have not been permanently damaged at all. They are perennials with stout subterranean rhizomes, and the following spring, the plants grow back from these rhizomes as dense and healthy as before. This beetle is not a biological control agent despite the very heavy feeding it does on this weed. In fact, the beetle makes the spatterdock a worse pest than it would otherwise be because it makes the plant so objectionable to look at in mid- to late summer. The plant has some esthetic appeal without the beetle, but with it, the plant has no redeeming qualities.

PROBLEMS WITH BIOLOGICAL CONTROL

The first problem with biological control of an aquatic weed then, is to find a control agent that will not become a pest itself; will not only survive, but thrive and reproduce in the new environment; and will not only attack, but actually control the weed. Finding such an organism is usually difficult and often impossible.

The second problem with biological control is that the level of control is sometimes cyclic, the pest going through periods of relatively low abundance followed by periods of relatively high abundance. This is illustrated graphically in Fig. 11.1. If we look at the left side of the graph we will see

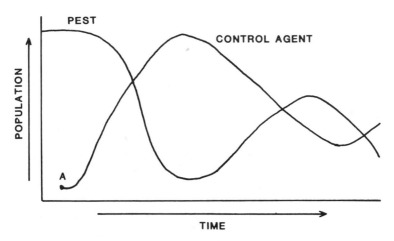

Fig. 11.1.   A representation of the cyclic nature which biological control sometimes assumes. A = time of introduction of control agent.

that the pest is abundant (at least abundant enough to be a problem or else it would not be a pest), and the control agent is at a very low level of abundance (because it was just introduced). As time goes on, the population of the control agent increases because of the abundance of its host or prey (the pest). Eventually, a point is reached at which the control agent begins to have an impact and the population of the pest begins to decline. This trend continues until finally the pest population becomes so low that it can no longer support the large population of control agents, which then declines either slowly or in a sudden collapse, depending on the situation. This, in turn, allows the pest population to begin increasing and eventually we have a situation similar to the one we started with—a high pest population and a low population of control agents. The cycle now begins all over again and may continue indefinitely, or the magnitude of the fluctuations may diminish with time until a kind of equilibrium is reached and the two populations remain relatively stable.

At the present time we have a poor understanding of the factors that determine whether or not such fluctuations will occur, and, if they do, how great the magnitude will be and how long they will continue. Any new biological control program runs some risk of going through these cycles.

The final problem with biological control is that while it may be inexpensive in the long run, it is very expensive in the short run. The cost of searching for a suitable control agent and then subjecting it to the rigorous testing procedures necessary to insure that it will be safe to introduce into the new environment is enormous. Approximately 12 years of research and testing are required, on the average, to develop a program for controlling a single weed species with an insect (Andres 1977). Further discussion of the nature of such testing and the associated costs will be found later in this chapter.

SPECIFIC EXAMPLES OF BIOLOGICAL CONTROL OF AQUATIC WEEDS

As recently as 20 years ago, there were almost no examples of successful biological control of weeds, and the few examples that existed all involved terrestrial species. During the past 20 years, however, a great deal of research has been done, and there are now probably more instances of successful biological control of aquatic weeds than of terrestrial weeds. Furthermore, research continues at a relatively high level and new programs using biological control agents against aquatic weeds are being undertaken at a fairly steady pace.

Arthropods as Control Agents

When one mentions biological control of a pest, most people probably think immediately of insects as being the control agents. This is understandable because so many of our successful programs have utilized insects in this capacity. They are common, they are present in enormous numbers, and they exist in what seems to be infinite variety. Furthermore, they reproduce rapidly and tend to spread quickly because of their ability to fly. All of these characteristics put them high on the list of potential agents for pest control.

The most successful program of biological control of aquatic weeds in the United States to date does in fact make use of an insect. It is the program for the control of alligator weed with the alligator weed flea beetle. Alligator weed is a serious pest of major economic importance that was introduced into this country sometime prior to 1900. It spread throughout the Southeastern States and California and in 1970 infested over 27,000 hectares (67,000 acres) (Bennett 1974). It is a very adaptable weed in the sense that it can grow as an emergent plant in water of moderate depth, it can grow as a semi-aquatic plant on wet shores or in low-lying, poorly drained areas, or it can grow as free-floating mats or "islands" in water of any depth. The plants that grow anchored to the bottom are particularly troublesome in irrigation ditches and drainage ditches, while those growing as floating islands interfere with navigation.

In 1959, a cooperative investigation was undertaken by the United States Department of Agriculture and the United States Army Corps of

Engineers to search for potential control agents for this plant. Excellent reviews of the nature of this search and the subsequent researching and release of the beetle are given in Spencer and Coulson (1976), Gangstad (1976), and Zeiger (1967).

Alligator weed is native to South America and the search was begun there, since it is logical to assume that natural enemies would have had more time to evolve in the plant's original range than anywhere else. A team of government scientists and technicians went to South America and began an extensive survey of the organisms found living on or in association with alligator weed. They found that the plant does not grow as vigorously or as abundantly in its native habitat as it does in the United States and other tropical and semi-tropical parts of the world into which it has been introduced. This was attributed, at least in part, to the action of between 40 and 50 species of insects that served as suppressing agents. Of these, four were determined to be major factors in controlling the plant. One was a flea beetle of the genus *Agasicles* that had never been described or named (Fig. 11.2). The others were a stem-boring moth, a thrips, and another flea beetle of the genus *Disonycha*.

Fig. 11.2.   The alligator weed flea beetle. The insect-damaged stem (left) shows adult beetles feeding on the outside and the split-open stem (right) shows pupae and larvae within.

In 1962 the decision was made to proceed with further testing of the new *Agasicles* species. (This insect was subsequently named *Agasicles hygrophila* and is commonly referred to as the alligator weed flea beetle). Testing was done at a laboratory set up specifically for that purpose at Castelar, Argentina. Extensive feeding tests at this laboratory indicated that the beetle could not complete its life cycle in the absence of alligator weed. In fact, only one other plant was found on which the beetle larvae would feed, and this was a plant very closely related to alligator weed. Although the larvae fed on this other species slightly, they wandered off of the plant after 2 or 3 days and died.

Here was an insect that appeared to have many of the necessary characteristics of a good biological control agent. All stages of the insect's life cycle occurred on the target species. The eggs are attached to the underside of the leaves of the host. The larvae and adults feed on the leaves and stems of the plant, and, when the larvae get older, they bore into the hollow stem where they feed for awhile and then pupate. The newly emerged adults bore to the outside and return to the leaves where they deposit more eggs and start the cycle over again. Furthermore, the insect was absolutely dependent upon the plant and was never found in the wild on any other species.

In 1963, after receiving the proper clearances from numerous federal and state agencies, the beetle was imported into the United States. It was shipped under strict quarantine to the United States Department of Agriculture Entomology Research Laboratory at Albany, California, where elaborate precautions were taken to insure that no insects escaped. Further studies and feeding tests were conducted at this facility to insure beyond all reasonable doubt that there was no danger in releasing the insect in this country.

Finally, after additional state and federal approvals, the first *Agasicles* were released on March 26, 1964, 5 years after the beginning of the search. This search and subsequent testing were as short and inexpensive as one might ever expect such a program to be because the organism was found and recognized as having potential very quickly. Also, the testing went quickly and smoothly because the organism was so ideally suited to the purpose that no serious questions about its safety ever arose.

Constant monitoring of beetle populations and alligator weed populations since 1964 have determined that the program has been a success. A qualified success, certainly, but a success nevertheless. The beetle has established itself in this country and has spread its range over much of the area occupied by alligator weed. In some areas it has effected excellent control of its host, but in other areas control has been marginal or unsatisfactory. The reasons for this lack of consistency are only partly understood, but they include climate and local habitat. Control is not good at either temperature extreme found within the range of alligator weed. Where winter temperatures are too cold, the host plant freezes back to the waterline and the beetles starve to death, having no diapause in their life cycle. On the other hand, long, hot summers reduce the fecundity of the beetle and prevent populations of sufficient density to control the weeds.

Two other factors have proved to be important in determining whether or not satisfactory control is obtained. One of these is the nutritional status of the host plant. In areas where alligator weed is well nourished, the beetle tends to thrive and to suppress the plant population. In areas where the weed shows symptoms of low nutrient availability, the beetle populations remain low and control is poor. The other factor is the moisture status of the site on which the plants are growing. The beetle is much more effective in controlling plants growing as true aquatics than those growing on wet shorelines or in low, swampy areas with little free-standing water.

In order to improve the situation, two other insects were subsequently introduced into the Southeastern States from South America. A thrips (*Amynothrips andersoni*) was introduced in 1967, and a stem-boring moth (*Vogtia malloi*) was introduced in 1971 (Spencer and Coulson 1976). Both have established themselves and work in conjunction with the beetle to give better overall control than the beetle was able to give alone. The moth is particularly effective since it is able to control the plant in climates where the winters are too cold for the beetle to be effective. It is also able to control the plant where it is growing as a semi-aquatic rather than as a true aquatic weed. The thrips has not been nearly as effective because it spreads very slowly and is subject to severe predation pressures in this country.

Although this program, combining the effects of three insects, is the most successful example of controlling an aquatic weed species with arthropods, it has not completely eliminated the need for other methods of control. Some mechanical control and some chemical control are still required. The impact of the insects can be estimated, however, from the following statistics regarding the cost of aquatic weed control at Semmes Lake, Fort Jackson, South Carolina. Prior to the release of the beetle and the moth, the average cost of labor, herbicide, and equipment rental for alligator weed control was $7,500 per year. Four years following the release of the beetle and the moth, the cost of additional control was $600 (Andres 1977). In other areas it has been shown that even though the insects do not provide adequate control of alligator weed by themselves, they weaken the plants enough so that the application of herbicides is far more effective than it would be in the absence of the insects. This is a perfect example of integrating biological control with chemical control.

Numerous other arthropods have been released in many tropical parts of the world in attempts to control aquatic weeds, but none has been so successful as the program on alligator weed in the United States. Among the others are an aquatic grasshopper for the control of water fern, and a number of insects and mites to control water hyacinth. The search for and testing of new arthropod control agents continues at a relatively high level of activity.

Snails as Control Agents

Two species of snails, both native to South America, have received serious attention as potential control agents for aquatic weeds. One of them, *Pomacea australis*, appears now to be unsuitable for use in this country, and research on it has virtually ceased. The other one, *Marisa cornuarietis* (Fig. 11.3), received considerably more attention by researchers and by the popular press during the 1960s, but it too has lost much of its support in recent years.

Marisa first came to the attention of the scientific community when it was discovered that it had the ability to control other snails that serve as intermediate hosts for the organism causing schistosomiasis in human beings. This fact was discovered after the accidental introduction of *Marisa* into Puerto Rico. Investigations began on the use of the snail for schistosomiasis control, and these investigations brought to light the animal's voracious feeding habits. In the early 1960s a flourishing colony was discovered in a canal near Miami, Florida. It is assumed that they were introduced there by tropical fish hobbyists who discarded them because of the damage they inflicted on ornamental aquarium plants. At one time they were widely sold in the tropical fish trade under the name "Columbian snails."

When their feeding habits became known, research began on their possible use as weed control agents (Seaman and Porterfield 1961). It was soon discovered that they would eat almost any form of organic matter including other snails, plants, and even newspaper when nothing else was available. In this sense, they were the exact opposite of the flea beetle described above, with its narrow, one-track diet. *Marisa* does have certain food preferences. When supplied with a variety of aquatic plants, it will feed on one or two favorites until they are gone before moving on to the next most favored species, but eventually, it will eat them all.

As a practical aquatic weed control agent, *Marisa* has some serious drawbacks. For one thing, it poses a definite hazard to aquatic crops such as rice (the seedlings are particularly vulnerable), watercress, and water chestnut. Another problem is that very large numbers of snails are neces-

Fig. 11.3.   *Marisa cornuarietis*. **The species of snail most seriously considered for use as a biological control agent for aquatic weeds in this country.** *Courtesy of Agricultural Research Service, USDA.*

sary to control dense growths of weeds. An estimate derived from one study was that it would take 20,000 snails per acre to control a dense stand of hydrilla. Once control is acheived, much smaller numbers are sufficient to keep the plants in check.

Another serious drawback is the fact that the lower lethal temperature for *Marisa* is 9°C (48°F). This precludes its use in all but the most tropical portions of this country. In order to overcome this drawback, it was once suggested that the snails be produced in enormous numbers in "snail factories." Persons in northern climates could then purchase the required number of snails in the spring of the year after the water temperature was sufficiently high, but before the weed population became too dense. The snails would then keep the weeds under control during the summer and would be allowed to die with the advent of cold water temperatures in the fall. If necessary, more snails would be purchased the following spring. This plan obviously requires that large numbers of snails can be produced in a year's time at a fairly reasonable cost. Some feasability studies were undertaken to investigate the potential of the plan but it never got beyond the earliest stages. At the present time, there is virtually no research being conducted in this country with *Marisa* or any other snail for weed control purposes.

Fish as Control Agents

Fish can control aquatic weeds in two distinctly different ways. They can eat the weeds or they can stir up the bottom mud continuously and to such an extent that the water remains turbid and shades out submersed vegetation. The latter conditions are sometimes created by catfish, carp, and certain minnows, but no one has vigorously pressed for the adoption of a program utilizing these fish for weed control. Among the many disadvantages are the facts that the constantly muddy water is a problem in itself, and the system will only work where the substrate is composed of mud.

The use of herbivorous fishes, however, has received widespread support and is being pursued avidly in many quarters. Although the role of plant-eating fish in vegetation control has been known for many years, particularly in the the Orient, attempts to manipulate fish populations specifically for aquatic weed control purposes have been restricted primarily to the past 30 or 35 years. Early research in this country involved two species of fish, both commonly known as silver dollar fish, and several species of the genus *Tilapia*.

The silver dollar fish (*Metynnis rooseselti* and *Mylosomma argenteum*) were investigated in the 1960s, but little follow-up work has been done. They are small fish from the tropical parts of South America that feed by biting off plants at the base of the stem near the soil surface. Floating mats of cut off vegetation result, which are grazed on by the fish but which are much too large for the fish to consume completely. The remainder usually blow into the shallows near the shore where they decay or occasionally root, starting new plants. Being tropical in origin, silver dollars do not graze actively below water temperatures of approximately 25°C (77°F) and their lower lethal death point is 16°C (61°F). This obviously restricts their use in any but the warmest parts of the world. Because of their small size, large numbers are needed in order to have an appreciable impact on a weed population. One estimate was that 500–1000 fish per acre would be a satisfactory number (Yeo 1967).

Another group of fish that has been investigated for its weed control potential is the genus *Tilapia*. Tilapias are widespread throughout Africa and the Middle East and are primarily tropical animals, although some species will tolerate cooler water than will the silver dollar fish. They are widely cultivated in their native range for human food, and their habits and requirements are fairly well known. Some species feed almost exclusively on algae, while others feed primarily on vascular plants. Next to their tropical temperature requirements, their biggest drawback is their enormous reproductive potential. They rapidly increase their numbers under favorable conditions and soon become the dominant species, crowding out other, less competitive fishes. One means of overcoming this problem is to stock them with an aggressive predatory species which can keep the tilapia under control.

A fact of interest is that the various species of tilapia can be interbred, creating the possibility of selective breeding to obtain a hybrid that has better than normal weed control characteristics and less tendency to overpopulate its habitat. It has already been shown that crosses of certain species result in offspring that are all males (National Academy of Sciences 1976). These all-male offspring could be raised in large numbers and could obviously be stocked without danger of overpopulation. Further investigation and hybridization could result in the tilapia becoming one of the major biological control agents for aquatic weeds on a worldwide basis.

One species, *Tilapia zillii*, is already being used in this country on an operational basis to control weeds in irrigation canals in the lower Sonoran desert of California (Legner and Pelsue 1980). Even in that very warm climate, however, overwintering survival is so poor that the fish must be restocked annually to maintain satisfactory levels of weed control. *Tilapia zillii* has a lower lethal temperature of 10°C (50°F), which limits its use to very warm parts of the world.

The fish that has the greatest potential for controlling aquatic weeds in nontropical parts of the world is, beyond all doubt, the grass carp, *Ctenopharyngodon idella*. This fish is a native of the Amur River system in northern China, and its natural range extends from there to southern China and into parts of Thailand. Since the mid 1950s the range of this fish has been extended by man to include much of Asia and North America and virtually all of Europe.

It was first introduced to North America in 1960 when it was imported into Mexico. Three years later it was introduced to the United States by the federal government at Stuttgart, Arkansas and by Auburn University at Auburn, Alabama (Sutton 1977). It was initially called "white amur" or "grass carp" in this country, but, by common usage, grass carp seems to have prevailed and the name white amur is being used less and less.

Fig. 11.4.   The grass carp. This fish is used in many parts of
the world as a source of food and as an aquatic weed control
agent. *Courtesy of Agricultural Research Service, USDA.*

The one characteristic of the grass carp that sets it apart from all other
biological control agents for submersed aquatic weeds is its tolerance of
low temperature. It is able to survive in water that is just above the
freezing point. In Arkansas, fish survived 5 weeks under a cover of solid
ice and then survived summer water temperatures of 35.6°C (96.1°F) with
no ill effects (Provine 1975). The upper lethal temperature for adults is
41°C (106°F), although very young fish may have a slightly lower maximum. This fish is clearly not restricted to the southern United States as are
the organisms discussed on the previous pages. Grass carp can probably
maintain themselves in every state in the United States and large parts of
Canada.

The grass carp tolerates wide ranges in other environmental parameters besides temperature. It can withstand dissolved oxygen levels as low as 0.3−0.5 ppm, although feeding ceases at approximately 2.5 ppm. These are exceedingly low levels for any fish. It can also tolerate more salinity than most freshwater fish. The exact amount of salinity it can tolerate is dependent on the age of the fish, the water temperature, and probably other factors, but there are records of adults surviving in diluted sea water with a salinity of 17.5 ppt. This may mean that the fish could migrate from one river system to another by passing through an estuary. They can also withstand unusually high pH values. Fish that were experimentally subjected to water having a pH of 10.8 showed no signs of stress and no ill effects.

The ability of the grass carp to control a wide variety of submersed and floating vegetation is beyond question. Even the most severe critics of the fish do not deny its ability to control weeds effectively. Unlike the silver dollar fish described above, which "mow" plants off at their bases, the grass carp feeds at the very tips of the leaves and stems and works its way down to the soil level. It is obligated to feed in this manner because the teeth it uses to grind the plant tissue are in its throat, and it is incapable of biting pieces of plant with its lips or the anterior part of its mouth.

The feeding habits of this fish have been studied in considerable detail. It was once claimed that the grass carp was completely herbivorous, but it is now known that it feeds on various kinds of zooplankton and minute aquatic insects when it is very small. As it approaches a length of 2 cm (0.75 in.), it begins to add macroscopic plants to its diet. Sometime between attaining a length of 4 and 5 cm (1.5 and 2.0 in.), it becomes solely herbivorous.

The grass carp will feed on a wide variety of plant species, although it prefers some types over others when given a choice. The degree of selectivity is, in part, a function of temperature. It feeds on the widest variety of plant species when the temperature is near the fish's optimum [above 27°C (80° F)], and becomes more selective as the temperature gets lower. As the temperature falls and the fish becomes more selective, it tends to stop eating the coarser, more fibrous species and concentrates on the softer, more succulent types. Feeding essentially stops below 12.2°C (54°F).

A peculiarity of the grass carp is that it has a digestive tract that morphologically resembles that of a carnivore. It is short, with a relatively straight, uncoiled intestine. As a result, it is very inefficient in digesting and assimilating the food it eats. Various studies have shown that the fish is able to digest and assimilate only between 20% and 60% of its food intake, depending on the nature of the diet and possibly on other factors.

From a practical point of view, this has both a negative and a positive impact on the use of the animal as a biological control agent. The negative aspect is that the feces of the fish contains a large amount of undigested or partially digested material, which decays in the water, thereby using dissolved oxygen and releasing dissolved plant nutrients that frequently result in an algae bloom. The positive aspect is that because of its digestive inefficiency, the fish has to eat a great deal more food than would otherwise be necessary, and in so doing, controls more plants than it otherwise would.

A major controversy surrounds the use of grass carp in the United States. This controversy revolves around the potential effect the fish might have on native fish populations. Data are available that show that under certain circumstances, at least, introduction of the grass carp has actually increased the numbers of other fish species. Equally valid data are available from other sources showing that the grass carp has had a negative impact on native fish species. Both the proponents and the opponents of releasing the grass carp are vocal in their arguments, and the issue is a long way from being settled to everyone's satisfaction. The grass carp has been widely released in Arkansas and several other states, but most states prohibit its importation or else will allow it to be brought in only under strict quarantine conditions for research purposes. Since young fish are available commercially and since the state laws banning importation are difficult to enforce, it is likely that the grass carp is already present in far more states than the official record would indicate.

The biggest danger in importing any new fish is that it will begin to reproduce and that enormous populations will build up, competing seriously with native fish, and possibly with ducks, muskrats, or other aquatic resources. For many years, the danger of this happening with the grass carp seemed remote because the conditions required for reproduction are so specific that it was felt they would never occur outside of the animal's native habitat. For natural reproduction to occur, the water level must rise, the temperature must rise, and the daylength must increase, all in the proper relation to one another. Furthermore, these things must happen in different ways for the males and for the females. Following spawning, the eggs must be transported in rapidly flowing water of high oxygen content and proper temperature. After hatching, the fry must have quiet water available to them with abundant zooplankton in order to develop. The chances of the fish finding all of these exacting conditions outside of its native range were felt to be so small that the danger of natural reproduction was almost nonexistant. In all the years that the fish had been in Europe, no natural reproduction has ever been reported. All breeding was done artifically by means of hormone injections. Both male and female fish receive injections of carp pituitary gland extract or human chorionic gonadotropin, which causes reproduction to occur.

Recently, however, the grass carp is reported to have spawned naturally in two separate locations in Mexico, where it was introduced not to control aquatic vegetation, but to serve as a source of protein for the local people (Sutton 1977). This report naturally supports the opponents of releasing any more grass carp in this country.

One means of insuring that no reproduction could occur in the wild would be to release only sterile fish. A step in this direction was taken in 1979 when hybridization was accomplished between female grass carp and male bighead carp (Hypophthalmichthys nobilis) (Sutton et al. 1981). Many of the offspring of this cross are triploids and should be incapable of reproducing. These hybrids are externally similar in appearance to their female parent and early indications are that their feeding habits are similar. Much more research is necessary to determine whether or not these hybrids will be useful and environmentally safe for general release.

One of the advantages of the grass carp that should not be overlooked is its potential as a food fish and as a sport fish. It may attain a length of 4 feet and a weight in excess of 100 pounds. Its flesh is not only edible but is considered to be excellent eating. In many countries it is looked on as a food fish first and a weed control agent second, if at all. It is rapidly replacing the common carp as the traditional Christmas Eve dinner in Poland. It is also a good sport fish, being sought with hook and line in many European countries. Willow leaves and other vegetation are used as bait although at least one report says the fish will also take artificial lures.

Whether we want it or not, the grass carp is probably in this country to stay. It remains to be seen whether it will prove to be a boon, a bane, or fall somewhere in between. The latter is most likely.

Mammals as Control Agents

A wide variety of mammals, ranging from moose to hippopotamuses to muskrats feed on aquatic vegetation, but only one, the manatee, has received serious attention as a potential biological control agent (Fig. 11.5). The West Indian manatee (Trichechus manatus), is one of four species of aquatic and marine mammals commonly called sea cows. They are large animals with fusiform bodies and nearly hairless skin. Their front limbs are modified into flipperlike or paddlelike appendages and hind limbs are lacking. The tail is modified into horizontal flukes, which are used in swimming. Manatees are confined exclusively to aquatic and marine habitats since they are completely helpless on land.

Fig. 11.5.   The West Indian manatee. This tropical mammal has received wide-spread publicity as an aquatic weed control agent, but in practice, has some serious drawbacks. *Courtesy of Denver Wildlife Research Center, USFWS.*

The West Indian manatee, like so many other organisms that have been proposed as biological control agents for aquatic weeds, is a tropical animal. It occurs in rivers, estuaries, and shallow coastal waters along the South Atlantic and Gulf coasts of the United States, the eastern coast of Central America, the islands of the West Indies, and the eastern coast of South America to a point slightly south of the Equator. In the United States, it is most abundant along the coasts of Florida. Even the Florida winters are at the limits of the manatee's temperature tolerance, however, and during the coldest months they congregate near sources of warm water such as constant-temperature springs. During the warmer parts of the year, they disperse along the coast.

In the late 1960s and early 1970s, the manatee was given a great deal of attention in the popular press as a solution to all of our aquatic weed problems. Its unique appearance, its large size, and its reputed role as the origin of the mermaid myths, made it an ideal subject for human-interest stories in newspapers and magazines. In fact, the scientific community was investigating the possibility of using them for aquatic weed control, and they were being successfully used for that purpose in canals in the South American country of Guyana.

As is invariably the case, the manatee has certain advantages and certain disadvantages as a biological control agent. Among its advantages are the facts that it consumes enormous quantities of plants, and it is an extremely safe control organism. It is harmless to man and his crops, and there is no chance at all of the population getting out of control and becoming overabundant.

As a matter of fact, the low population densities and the failure to increase its population is the greatest drawback to use of the manatee. It is a rare animal, which is on the endangered species list. Although it is protected by law in almost every country in which it occurs, the laws are difficult to enforce, particularly in remote areas of South and Central America and the more remote areas of the West Indian islands. The manatee is widely hunted for its meat, its hide, its oil, and its ribs, which are polished and used as a substitute for ivory. Even in more populated areas where law enforcement is more effective, manatees are frequently shot for "sport," although the sluggish, docile manatee is certainly not a challenge as a game animal.

Another problem is that large numbers of manatees are killed each year wherever manatees and motorboats occur together. The vast majority are accidental. Manatees have the unfortunate habit of basking just beneath the surface of the water where they cannot be seen by boaters. When a boat hits one, the keel rides up over the manatee's back and slides along until the propeller hits the animal, frequently killing it immediately by severing the spinal column. Many of Florida's manatees have scars on their backs resulting from nonfatal encounters with motorboats.

Yet another factor that contributes to the small size of the manatee population is its exceedingly low reproductive capacity. Each cow has a single calf at a time and never more frequently than once every 3 years. Until very recently, it was thought that twins never occurred, but it is now believed that it does happen on rare occasions. The gestation period appears to be somewhat variable in length but frequently exceeds 13 months. Maternal care lasts for a year or more. With such a low reproductive rate, it is very difficult to increase populations even in captivity or semi-captivity where poaching and accidental deaths are held to a minimum.

They fare poorly in captivity and frequently succumb to a respiratory infection that is somewhat similar to pneumonia in humans. Most captive specimens require frequent injections with antibiotics to remain healthy. Captives in Guyana appeared to have been exceptions, remaining healthy for many years (National Academy of Sciences 1976).

Although manatees will eat a wide variety of aquatic plants, they have definite preferences and will often travel long distances to feed on a favorite species, bypassing other, less favored species along the way. For this reason, their greatest potential is in small bodies of water or canals where they are confined to a reasonably small area and cannot leave to look for other kinds of food. Where manatees were successfully used for aquatic weed control in Guyana, they were confined to man-made canals and ponds.

At the present time the manatee is not thought to be a viable candidate for widespread use as a biological weed control agent. This is primarily due to its status as an endangered species, its scarcity, and its low reproductive capacity.

Plants as Control Agents

Plants, both macroscopic and microscopic, have been used to control aquatic weeds. They do this through competition for light, nutrients, or space, or they do it by a process known as allelopathy. Allelopathy is the release into the environment by one species of plant of a chemical or chemicals that suppress or kill other species of plants.

Control by competition has been the method most frequently employed. The earliest and most widespread method of controlling aquatic weeds with other plants has been to shade them with dense blooms of planktonic algae. In the Southeastern United States, it is common practice to apply inorganic fertilizers to fish ponds in order to induce an algae bloom. This increases fish production by increasing primary productivity and thereby increasing the carrying capacity at each trophic level. A second reason for inducing an algae bloom is to control submersed aquatic weeds. The primary disadvantage of this method is that the dense population of algae makes the pond unsuitable for many uses, such as swimming. There is also the danger of fish suffocating under conditions of darkness, as explained in Chapter 8.

The use of vascular plants as competitors has been looked into only recently and to a much lesser extent than the use of algae. The most prominent and successful example to date of the intentional use of one species of vascular aquatic plant to control another is the use of *Eleocharis coloradoensis*, a low-growing, submersed, spike rush. This plant has proved to be very effective in controlling various species of pondweeds that interfere with the flow of water in irrigation canals in California (Yeo 1980). The spike rush forms a dense sod on the bottom of the earthen canals that prevents the colonization of the area by pondweed or even replaces existing stands of pondweed. The spike rush itself does not interfere with the flow of water because of its low growth. The presence of spike rush has certain advantages over an unvegetated bottom. For example, it helps to prevent erosion, and it provides suitable habitat for invertebrates that serve as fish-food organisms.

The use of this plant has not become widespread outside of the canals described above. At least part of the explanation for this is that the plant thrives best when it is exposed to the air for a brief time each year. The canals in which it is used in California are dewatered for a few months each year as a part of their regular cycle of use, and this dewatering comes at the proper time and lasts for the proper duration to benefit the plants.

Although no work has been done to date, a promising avenue of research would seem to be the breeding of special strains or varieties of low-growing, highly competitive plants that are adapted to various climates and habitats. They would have none of the disadvantages of aquatic weeds but would be far more desirable than a totally unvegetated bottom.

Virtually no work has been done on allelopathy in aquatic species, although some data have been published suggesting that cattail may release allelopathic substances, at least under certain conditions (Szczepanski 1977). It is possible that the spike rush described above releases a chemical compound into the environment that suppresses the pondweeds it controls so successfully. One would assume that allelopathic compounds would be diluted very quickly in an underwater environment by diffusion and by the action of currents. Root exudates released into the hydrosoil, might remain in place longer and be more effective than those released into the water above the soil.

Plant Pathogens as Control Agents

Aquatic plants are subject to diseases just as terrestrial plants are. The concept of using the causitive agents of these diseases to control unwanted plant populations is very appealing but work in this area is in its infancy and operational, field-scale use of pathogens has never been successful. A wide variety of viruses, bacteria, and fungi have been isolated from diseased aquatic plants that were collected in the field, and some of these pathogens have been cultured on a reasonably large scale in the laboratory. They have been used to inoculate plants in the field on an experimental scale with varying degrees of success (Freeman 1977).

The weed species receiving the most attention in regard to pathogens is water hyacinth. Several promising fungi have been found that infect this plant. All have drawbacks of one kind or another, but active programs are underway to find new pathogens and to evaluate further those already known. Other research is being conducted to investigate the possibility of using two or more pathogens in combination or a pathogen in combination with an insect or other arthropod.

There are at least two instances in which extensive natural reductions in aquatic plant populations have been attributed to pathogens. One instance was the dramatic decline of Eurasian water milfoil in Chesapeake Bay in the 1960s (Elser 1969). This has been attributed to a virus, although it has never been proved to everyone's satisfaction. It is possible that natural environmental changes (particularly salinity) also played a role in the decline. It is not beyond reason to believe that environmental changes made the plants less resistent to the virus and allowed the disease to erupt in epidemic proportions.

The most frequently cited example of the decimation of an aquatic plant population by a pathogen occurred in 1931 and 1932, when a catastrophic decline occurred in the populations of eelgrass along both the American and European coasts of the North Atlantic Ocean (Rasmussen 1977). Eelgrass is of enormous ecological importance in the shallow marine waters where it occurs, and the epidemic was widely studied in America and Europe. The pathogenic condition was referred to as the "wasting disease" of eelgrass, and although complete agreement was never reached on an explanation for the disease, the most commonly agreed-upon cause was a slime mold and a fungus infecting the plant at the same time. Whatever the cause, the result was devastating to the eelgrass. Twenty years later, the plant had still not recovered to its former abundance, nor had it recolonized all of the areas it had once inhabited.

Other Organisms as Biological Control Agents

Numerous other organisms have been mentioned briefly in the literature as possible biological control agents. Among them are ducks, geese, swans, crayfish, water buffalo, and nutria. None have been widely used or researched, and none is likely to become a major means of aquatic weed control in the foreseeable future. This is not to say, however, that some of these organisms may not be useful in local situations under specific conditions. All of the animals mentioned above are useful to man for other purposes, and in localities where they are being raised for food, fur, or as draft animals, aquatic weed control may be a secondary benefit.

BIBLIOGRAPHY

ANDRES, L.A.   1977.   The economics of biological control of weeds. Aquatic Bot. **3**, 111–123.

BENNETT, F.D.   1974.   Biological control. *In* Aquatic Vegetation and Its Use and Control. D.S. Mitchell (Editor). United Nations Educational, Scientific and Cultural Org., Paris.

ELSER H.J.   1969.   Observations on the decline of the water milfoil and other aquatic plants, Maryland 1962-1967. Hyacinth Control J. **8**, 52–60.

FREEMAN, T.E.   1977.   Biological control of aquatic weeds with plant pathogens. Aquatic Bot. **3**, 175–184.

GANGSTAD, E.O.   1976.   Biological control operations in alligator weed. J. Aquatic Plant Management. **14**, 50–53.

LEGNER, E.F., and PELSUE, F.W., Jr.   1980.   Bioconversion: Tilapia fish turn insects and weeds into edible protein. California Agric. **34** (11), 13–14.

NATIONAL ACADEMY OF SCIENCES.   1976.   Making Aquatic Weeds Useful: Some Perspectives for Developing Countries, Washington, DC.

PROVINE, W.C.   1975.   The Grass Carp. Texas Parks and Wildlife Dept. Austin, TX.

RASMUSSEN, E.   1977.   The wasting disease of eelgrass *(Zostera marina)* and its effects on environmental factors and fauna. *In* Seagrass Ecosystems, A Scientific Perspective. C.P. McRoy and C. Helfferich (Editors). Marcel Dekker, New York.

SEAMAN D.E., and PORTERFIELD, W.A.   1961.   Control of aquatic weeds by the snail *Marisa cornuarietis*. Weeds **12**, 87–92.

SPENCER N.R., and COULSON, J.R.   1976.   The biological control of alligator weed, *Alternanthera philoxeroides*, in the United States of America. Aquatic Bot. **2**, 177–190.

SUTTON, D.L.   1977.   Grass carp *(Ctenopharyngodon idella* Val.) in North America. Aquatic Bot. **3**, 157–164.

SUTTON, D.L., STANLEY, J.G., and MILEY, W.W., II.   1981. Grass carp hybridization and observations of a grass carp X bighead hybrid. J. Aquatic Plant Management **19**, 37–39.

SZCZEPANSKI, A.J.   1977.   Allelopathy as a means of biological control of water weeds. Aquatic Bot. **3**, 193–197.

YEO, R.R.   1967.   Silver dollar fish for biological control of submersed aquatic weeds. Weeds **15**, 27–31.

ZEIGER, C.F. 1967. Biological control of alligatorweed with *Agasicles n. sp.* in Florida. Hyacinth Control J. **6**, 31-34.

12

Chemical Control of Aquatic Vegetation

Chemical weed control is the use of a chemical compound to kill weeds, prevent the germination of weed seeds, or rarely, to modify the growth of the weeds in some manner so that they do not pose as great a problem as they did before the chemical was applied. The chemical compound used for controlling the weeds is called an herbicide, or, if it modifies the growth of the plant, it may be called a plant growth regulator.

The only type of chemicals used to control aquatic weeds on an operational scale have been herbicides that kill plants after germination. Germination inhibitors and plant growth regulators that change the form of growth have been used experimentally but with limited success. Interest in plant growth regulators for use on aquatic weeds has not completely disappeared, and some research continues along these lines.

The use of chemicals to control unwanted plant growths in water is not new. Copper sulfate was successfully employed to control algae in 1904 (Moore and Kellerman 1904) and sodium arsenite was used to control vascular aquatic plants at least since the early 1900s. According to one source (Younger 1965), the earliest commercial application of an herbicide to a lake by a professional applicator occurred in 1924. The use of herbicides in aquatic areas by private individuals certainly preceded that date by a decade or more.

Until the discovery of the herbicidal properties of 2,4-D in the 1940s, all herbicides in common use were inorganic compounds. The discovery and widespread use of 2,4-D after World War II ushered in a new era in weed control. Thousands of substances were screened annually in an attempt to find new chemicals with herbicidal properties. Numerous chemical companies, governmental agencies, and universities were engaged in the search, and scores of new herbicides—virtually all of them organic compounds—were discovered. While the primary goal of the searchers was to find better herbicides for use in cultivated crops, non-crop areas were not ignored. Considerable effort was directed at the development of suitable weed control chemicals for roadsides, railroad rights of way, industrial areas, and aquatic sites. During the 1950s and early 1960s several new herbicides were registered with the United States Department of Agriculture for use on aquatic weeds. Among them were 2,4-D, silvex, diquat, and endothall.

It should be pointed out that every commercially available herbicide has at least three names. It has one or more trade names, which are usually registered trade marks of the company or companies manufacturing, formulating, and/or distributing the product. The basic herbicide may be sold under different trade names, and the products may be identical, or they may vary in their formulations or in their concentrations of active ingredients. Trade names will not be used in this book.

The second type of name is a common or generic name. These names are approved by the Weed Science Society of America and are comparable to the generic names of drugs. They will be used here unless otherwise specified. The generic name applies to the active ingredient only. The active ingredient is that substance in the commercial product that actually exerts the herbicidal effect on the plant. The herbicide names used above (2,4-D, silvex, diquat, and endothall) are all generic names. Each refers to a specific chemical compound that may be present in a single commercial formulation or in several commercial formulations whose other ingredients vary. These other ingredients may be solvents, diluents, surfactants, etc.

The third type of name that each herbicide has is its chemical name. It is derived by employing the nomenclature system used to name all organic compounds based on their molecular structure. Thus, the chemical name of endothal is 7-oxabicyclo[2.2.1]heptane-2, 3-dicarboxylic acid, and the chemical name of 2,4-D is (2,4-dichlorophenoxy)acetic acid. Chemical names of organic herbicides will not be used in the body of this book, but a list of generic names along with the corresponding chemical names appears in the Appendix.

It was mentioned that several herbicides were registered with the United States Department of Agriculture in the 1950s and 1960s. Under the provisions of the Federal Insecticide, Fungicide, and Rodenticide Act, with its several amendments, all pesticides sold or used in the United States must be registered with the federal government. From the time of the passage of this act in 1947 until 1972, responsibility for registration lay with the Department of Agriculture. Following the creation of the Environmental Protection Agency (EPA) in 1972, responsibility for pesticide registration was transferred to that agency. Without EPA registration, no pesticide may be legally offered for sale, sold, or used in this country.

Before such a registration is granted, the applicant must provide EPA with a great deal of information about the herbicide and its behavior in the environment. The applicant is almost always the company that intends to produce and/or market the chemical. In addition, the company needs certain basic information about the compound for its own use, even if it is not required by EPA for registration purposes. Let us look at a hypothetical new compound that has just been synthesized for the first time, and see what kind of information has to be obtained before it can be registered and sold for use as an aquatic herbicide.

INFORMATION REQUIRED FOR REGISTRATION

Preliminary Information

The first and most obvious question pertains to the new compound's efficacy. Will it or will it not control weeds? If not, no further information is required because it obviously has no value as an herbicide. Preliminary information is determined by the company in a primary screening test to which thousands of chemicals are subjected each year. Assuming that our hypothetical compound gets through the primary screening test by exhibiting some degree of herbicidal activity, we now have to obtain a somewhat better idea of what dosage rates are necessary to kill plants and perhaps a slightly better understanding of the spectrum of plants that are affected. This information, still very preliminary in nature, is obtained from a secondary screening test, which utilizes a wider variety of plants and dosage rates than the primary screening test.

If the new compound successfully passes the secondary screening test, it will then be subjected to numerous other tests and to scrutiny by experts in the fields of law, economics, chemistry, biology, engineering, and marketing. These tests and analyses do not always follow one another in exactly the same order and many of them are run concurrently. They are all designed to answer the questions that are considered in the following paragraphs.

Safety

How safe is the material? This is a question in which EPA and the general public are vitally interested. Toxicity to warm-blooded animals must be determined. This includes both chronic and acute toxicity, and tests must be conducted on various species of birds and mammals. Safety to humans is not the only consideration. Safety to wildlife, livestock, pets, and invertebrate animals must also be considered. Nor is direct toxicity the only concern. Subtler effects such as teratogenicity must be tested for, as well as effects such as the well-publicized thinning of egg shells in birds. Safety to irrigated crops must also be investigated. If the candidate herbicide is to be used in water, it is imperative to know what effect it may have on crops that are irrigated with that water. This includes not only commercial crops but home lawns, shrubbery, and other ornamental plantings.

Efficacy

How effective is the herbicide and what factors affect its efficacy? The early screening tests told us that the material exhibited some herbicidal properties and gave us a very general idea of what rates of application were necessary in to order obtain phytotoxicity. Now we need numerous, detailed, replicated tests to find out precisely how much herbicide is needed under different conditions. Is the efficacy affected by water chemistry? Is it affected by the age of the plant? Is it affected by temperature, sunlight, or the physiological state of the plant? These and a host of other questions must be answered before our candidate herbicide can become a commercial, legally available herbicide. In order to answer these questions it is necessary that field testing be done over a wide geographic area under diverse environmental conditions. These field tests are conducted under carefully controlled and monitered conditions by personnel of chemical companies, various state and private experiment stations, and the federal government. The final result of this field testing program should be precise knowledge of how much herbicide need be applied to control a specific type of weed under a given set of environmental conditions, and when and how the herbicide should be applied.

Environmental Fate

The next question is, "What is the fate of the compound when introduced into an aquatic environment?" Before the Environmental Protection Agency will register any aquatic herbicide, it must know what happens to the material after it is put into the water. What is its solubility under different conditions? What might cause it to precipitate? Will it vaporize and get into the air? Is the molecule degraded by ultraviolet light? If so, does enough ultraviolet light from the sun penetrate the water's surface to have a significant effect? Does the molecule undergo biodegradation within the tissues of animals or plants? If the compound is degraded in some way, what is the nature of the degradation products and what is their fate? Are the molecules adsorbed onto the surfaces of soil particles and organic debris? If so, how tightly are they bound?

A question of utmost importance is, "Does the chemical accumulate in the bodies of fish and other aquatic animals?" If so, in what tissues do they accumulate? Are these the tissues normally eaten by humans, such as muscle tissue, or are they tissues normally discarded by humans such as brain tissue?

Another consideration is the matter of biological magnification. This is a phenomenon that occurs when animals store a particular chemical in their bodies and do not excrete it at all or excrete it very slowly. The concentration of the chemical in the animal then increases over a period of time. Furthermore, the concentration increases with each trophic level in the food chain because each level is ingesting animals that have already concentrated the chemical in their own bodies. Thus, the higher one goes on the food chain, the more concentrated the chemical becomes. If man is the ultimate consumer, he can ingest significant amounts of the chemical. Even if man is not directly involved as a consumer, it is important that we know whether or not biological magnification is occurring and what the effects might be on the organisms involved.

Ultimately, we must find out exactly what happens to the herbicide added to the water of a lake, pond, estuary, or stream. We must know where it goes, what form or forms it will take, and what effects it might have.

Spectrum of Activity

The spectrum of activity of an herbicide refers to the variety of plants the herbicide affects. Another closely related term is "selectivity." It is a well-established fact that all herbicides do not affect all species of plants in the same way or to the same degree. At one extreme there are a few commercially available herbicides that will kill virtually any plant with which they come in contact. None of them are used in aquatic areas. At the other extreme are highly selective materials that affect only a few families of plants, or, in a few cases, are even able to control weeds that are in the same botanical family as the crop on which they are used.

It is imperative that we find out which plants are affected and which ones are not, but it is even more important that we find out *why* some plants are affected and some are not. This basic understanding will make it possible for us to use the herbicide in a much more intelligent manner in a much wider range of conditions. It will allow us to predict what the selectivity will be in advance. This is important because it is obviously impossible to test an herbicide for effectiveness against every species of plant. It is always better to know why something happens than to know merely that it does happen.

Selectivity is the result of various factors working alone or in combination to cause various plant species to respond differently to the same herbicide. One of these factors is the biochemistry of the plants themselves. One species of plant may contain enzymes that detoxify the herbicide while another species of plant does not. This is a common cause of selectivity. More rarely, the herbicide itself, as it comes from the can or bag, is not toxic but certain species of plants have the ability to convert it into a phytotoxic compound. In this situation, the susceptible plants are the ones that alter the herbicide, and the resistant plants do not.

Selectivity may also be the result of differential uptake by different species of plants. The herbicide may be equally toxic to two species of weeds if it is within the plant tissues, but because of a waxy cuticle, a hairy leaf surface, or some other obstacle, the herbicide is able to enter one plant and not the other. This is a form of physical selectivity.

Selectivity is not always absolute. It is common for the relative susceptibility or resistance of a plant to a particular herbicide to vary with rate of application, stage of growth of the plant, or physiological state of the plant as affected by temperature, sunlight, and even humidity. It is thus possible in some situations to control a particular species of plant or to leave it unaffected by choosing the proper rate of herbicide application, the proper timing of application, and sometimes by choosing the proper weather conditions.

Mode of Action

The mode of action of an herbicide refers to the specific means by which the active ingredient kills or alters the plant after the molecules are within the tissues. Which specific biochemical reactions are blocked or initiated that cause the plant to react in the way it does? For example, if the plant dies because it can no longer manufacture chlorophyll, we must find out which specific step in chlorophyll synthesis is blocked. Some herbicides affect more than one biochemical reaction; others are very specific, affecting a single reaction at a single site.

Formulation Effects

This involves the chemical or physical formulation of the commercial preparation of the active ingredient. You will recall that a commercial herbicide contains an active ingredient that exerts the herbicidal effect, but it also contains various other materials such as solvents, fillers, and surfactants. The nature and amount of these other ingredients may very well have an effect on the efficacy, fate, or spectrum of activity of the commercial preparation.

An herbicide can be formulated in many different ways. Its physical state may be a liquid, a dry granular preparation, a soluble powder, or an emulsion. Other, less common preparations have also been used. In the case of liquids, it is easy to understand that the solvent may have an effect by having some degree of phytotoxicity itself, or by affecting the rate of absorption of the active ingredient into the plant. In the case of granular preparations, the nature of the granule is important. Will the granules float, releasing the active ingredient at the surface of the water or will they sink, releasing the active ingredient at the bottom? Will the active ingredient be released over a period of minutes, hours, or days? This is a vital question, especially if the water is flowing even slightly. Granules have been made of many different materials including ground corn cobs, crushed walnut shells, and plastics, but the only really common material is clay. By choosing the proper type of clay and baking the granules at the proper temperature for the proper length of time, the rate of release of the impregnated active ingredient can be controlled.

In addition to the physical formulation of the herbicide, there is the question of the specific chemical form the active ingredient should have. The same basic active ingredient may be present in the commercial preparation as an acid, for example, or as the ester of that acid, the amine of that acid, the sodium salt of that acid, etc. Each chemical form may have slightly different effects on plants or behave differently in aquatic environments.

All of these physical and chemical formulations must be considered and many of them must be made and tested in order to make sure that the final product is as safe, as economical, and as effective as possible.

Patent Position

A question of prime importance to the company developing the herbicide but of little or no interest to the Environmental Protection Agency is, "What types of patent protection are available for this new chemical?" Can the molecule itself be patented? Can a use patent be obtained on its weed-killing properties? Can the manufacturing process be patented? If the answer to all these questions is "no," then it is highly unlikely that the herbicide will ever be developed and made available for use. Although we will not go into any detail on the matter of patents and patentability, it is a crucial test which every newly synthesized compound must pass before it can become a commercially available product.

Financial Considerations

Before our hypothetical new chemical advances very far through its progression of chemical and biological tests, experts will begin to examine the financial aspects of its development and sale. Market analysts will make estimates of the potential market for such a product. They will try to determine how many acres of water are infested with the specific types of weeds for which this herbicide appears to be effective. They will try to determine the geographic distribution of the potential market. Is the market concentrated so that distribution costs will be minimal or is it widespread so that distribution costs will be high? They will also try to determine whether the average individual buyer will need small quantities or large quantities because that will affect the packaging of the product, which in turn will affect the price structure. The pricing of a small-package market is very different from that of a bulk market.

Other financial considerations are the presence or absence of competitive products and their effectiveness, estimated manufacturing costs, and the possibility of finding other uses for the chemical, which would help to increase sales.

It is obvious that a great deal of time and money are necessary to synthesize a new compound and to develop it, register it, and bring it to market. The cost and time required have risen dramatically in the past fifteen years. Table 12.1 shows the estimated cost and time for several selected years. Costs have risen as the federal government has required more and more research data prior to the granting of a registration.

Because of the large financial investment required to develop and register a new aquatic herbicide, and considering the size and nature of the potential market, there is relatively little developmental effort being exerted along these lines by the chemical industry at the present time. The general feeling is that the chances of producing a marketable new aquatic herbicide and the potential profit to be made in the event that one is produced, do not warrant the financial investment required. As a result, most of the aquatic herbicides that are registered and commercially available at the present time are older compounds that have been available for a number of years. It is unlikely that a large number of new materials will be registered in the near future.

TABLE 12.1.   Approximate Time and Cost Required for the Development and Registration of a New Pesticide

Year	Approximate cost (Millions of $)	Approximate time (Months)
1967	3.4	60
1970	5.5	77
1976	12.0	84
1980	15.0	90
1981	20.0	96
1982	29.0	120
1983	37.0[a]	138[a]

[a] Estimated.

SPECIFIC HERBICIDES

Below are brief discussions of the most commonly used aquatic herbicides in the United States today. It should be understood that there are other registered herbicides that are less frequently used and still others, which, under special provisions of the Federal Insecticide, Fungicide, and Rodenticide Act, are registered for use in a limited geographic area only. Still others are in widespread use outside of the United States although they cannot be used in this country at all.

Copper Sulfate

Copper sulfate is the only inorganic herbicide in general use in water. It is 100% active ingredient and is available in various particle sizes from a very fine powder called "snow," to pieces that are several inches in diameter. The finer particle sizes are best because they dissolve more easily. Within the rates of application recommended for herbicidal use, copper sulfate is effective against *algae only*. It is not effective against vascular aquatic plants at application rates that are safe for fish and other aquatic organisms.

The toxicity to warm-blooded animals is low, but the powder should not be allowed to enter the eyes or come in contact with other mucous membranes. Treated water may be used for human consumption, irrigation, stock watering, and all forms of recreation, immediately after treatment with copper sulfate.

Other copper-containing herbicides that have the copper atoms bound in various chelates and organic complexes are also on the market. Their primary advantage over copper sulfate is that the copper does not precipitate out of solution as rapidly after it is applied to the lake, pond, or stream. These preparations are especially useful in hard water situations where copper has a tendency to precipitate out of solution very rapidly, making it unavailable to the algae. None of them are useful for controlling vascular plants or mosses.

2,4-D

This is the oldest organic herbicide in use today and the first organic herbicide to be used in aquatic situations. It was used to control Eurasian water milfoil in New Jersey lakes and Chesapeake Bay in the late 1950s. It is manufactured and sold under a wide variety of trade names and in many different physical and chemical formulations. A special granular formulation containing 20% 2,4-D (by weight) is produced specifically for aquatic use. No other formulation should be used for submersed aquatic plants, although liquid formulations may be sprayed on emergent shoreline vegetation. Water pH may influence the effectiveness of this herbicide. Acid conditions (pH below 6.0) tend to enhance the herbicidal action of the chemical, while alkaline conditions (pH above 8.0) tend to reduce the herbicidal action. Water treated with 2,4-D should not be used for drinking, livestock watering, irrigation, or agricultural sprays. All recreational uses of the water may proceed immediately after treatment.

Endothall

Endothall is available as either a liquid or granular preparation and in several different chemical forms. All are registered for aquatic use. Necessary waiting times required before using the treated water for various purposes such as drinking, livestock watering, and irrigation vary with the formulation and proposed water use. Each has its own waiting times and precautions that must be observed. Endothall is a wide-spectrum herbicide that will control the majority of troublesome submersed aquatic weeds. It is highly irritating to mucous membranes, and precautions should be taken to keep it out of eyes, nose, mouth, and other sensitive areas.

Diquat

Diquat is available only as a liquid containing 2 pounds of active ingredient per gallon. It is used in much the same manner as endothall and controls roughly the same spectrum of weeds. Irrigation, swimming, and livestock watering may proceed 10 days after treatment. Do not use treated water for human consumption until 14 days after treatment. A peculiarity of diquat is that it is adsorbed very tightly onto the surfaces of clay particles and organic detritus. It cannot be taken into living plants when it is thus adsorbed, so it should not be used in excessively muddy waters or when plant surfaces are covered with silt.

Simazine

Simazine is sold as a white powder to be mixed with water to form a suspension prior to application. It is sold under several trade names, but only one of them is registered for aquatic use. It is useful in controlling both algae and certain vascular plants, but it is very slow acting, particularly on the vascular plants. Treated ponds may be used for swimming and fishing immediately after treatment, but water for irrigation, stock watering, and human consumption should not be drawn for 12 months following treatment.

Dichlobenil

This herbicide is available as a granular material for aquatic use. It cannot be used in potable water supplies or in water to be used for irrigation or for livestock watering. Fish from treated waters should not be used for human consumption or for animal feed for a period of 90 days following application. These restrictions severely limit the use of dichlobenil. It will control a relatively wide spectrum of weeds but is especially useful for the macroscopic algae of the family Characeae, which are resistant to many of the other registered compounds.

Fenac

Fenac is unique in that it is the only currently registered herbicide for the control of aquatic weeds that is not applied to the water or to the foliage of the weeds. The lake or pond to be treated is first drained and the fenac is then applied to the exposed bottom. This, of course, requires that the body of water to be treated is physically capable of being drained and that the draining will not interfere with any vital use of the lake, pond, or reservoir.

Dalapon

This material is not registered for use directly in water but is useful for controlling grassy weeds such as giant reed in wet, low-lying areas and on the banks of ditches, ponds, and streams. It is available as a soluble powder, which is dissolved in water prior to application.

Amitrole

Amitrole is similar to dalapon in that it cannot be applied to water. It, too, is used on shoreline weeds and those growing on the banks of drainage ditches, canals, and in similar locations where emergent "semi-aquatic" vegetation needs to be controlled. It has a broad spectrum of activity and unlike dalapon, its use need not be restricted to grasses.

CHOOSING AND APPLYING HERBICIDES

The choice of herbicide in any particular situation is governed primarily by two factors: the species of weeds to be controlled and the uses being made of the water. The very first step in any aquatic weed control program is the proper identification of the weeds. Without proper identification of the plants to be controlled it is impossible to proceed with the intelligent selection of an herbicide. After the weeds have been identified, it is necessary to find an herbicide that will control the species causing the greatest problems. It may not always be possible to control all of the problem species with a single chemical.

Help in selecting the best possible herbicide is available from several sources. The Cooperative Extension Service in each state has people available at the county level to assist in making the choice. The Land Grant Colleges in most states have printed recommendations on the herbicides of choice for important aquatic weed species in their own states. State fish and wildlife authorities may also provide assistance with the selection of appropriate herbicides in some instances. Pesticide dealers, too, can be a source of help.

After a list of potential herbicides has been compiled based on the weeds to be controlled, the next step is to consider the uses being made of the water and to compare them to the use restrictions placed on each of the chemicals. These restrictions are generally available from all of the sources listed above and are always clearly stated on the herbicide label. If the use of a particular herbicide requires that there be no irrigation with treated water for 90 days after treatment, and if this is an unacceptable condition, then it is clearly not possible to use this material. To use an aquatic herbicide (or any pesticide) in a manner not in accordance with the provisions printed on the label is not only foolish, it is a violation of the law.

Every pesticide label is a legal document in the sense that the information on that label is approved as a part of the registration process. All of the directions for use, precautions to be taken during and after application, rates of application, and similar information, have been arrived at through the long, expensive developmental process described earlier, and have been carefully considered and approved by the Environmental Protection Agency. The law requires that the material not be used in any manner except that stated on the label.

When the list of possible herbicides to be used has been narrowed down to those that are effective against the weed species in question and those compatible with the use of the water, other criteria such as cost, convenience, and local availability are used to make the final decision.

The timing of aquatic herbicide applications, while not absolutely critical, is, nevertheless, important. As a general statement, herbicides should be applied in the spring after the weeds have begun to grow vigorously but before the submersed species have reached the surface and begun to form thick floating mats. Waiting until later in the season can cause any of several problems to occur. Dense stands of weeds can result in poor dispersion of the herbicide through the water. They can also interfere with the movement of boats, resulting in poor distribution of the herbicide. Furthermore, some herbicides are more effective when the weeds are physiologically active, as many of them are during the spring growth period.

The most important reason for treating in the spring, however, is to minimize the chances of suffocating fish. Whenever a large mass of organic matter decomposes in a body of water, dissolved oxygen levels in the water are depressed. The greater the mass of organic matter and the more rapidly it decomposes, the greater will be the depression in dissolved oxygen levels. Dead aquatic weeds comprise just such a mass of organic matter and waiting until later in the summer to treat means that a greater mass of weeds will probably be present to die and decompose. Waiting until later in the summer also means warmer water temperatures. Warmer temperatures increase the rate of decomposition, which in turn depresses the dissolved oxygen levels further and increases the risk of suffocating fish. In addition, warm water is not capable of holding as much dissolved oxygen as cool water, so that even before the weeds begin to decompose, oxygen levels will be lower in the summer than in early spring, adding to the danger.

The equipment used to apply the herbicide to the water varies with the physical form of the material to be applied and with the size of the area to be treated. Small ponds are frequently treated with hand-held, garden-type sprayers (Fig. 12.1) or small rotary granule spreaders, which hang around the applicator's neck and disperse the granules from a spinning disk actuated by a hand crank (Fig. 12.2). At the other extreme are very large areas, which are treated with heavy industrial-type equipment including helicopters, fixed-wing aircraft, and barges and spray-boats built specifically for the purpose of applying aquatic herbicides. The Tennessee Valley Authority treats thousands of acres of reservoirs each year for the control of Eurasian water milfoil. They have large barges that serve as floating heliports and re-supply stations for helicopters that spread granular 2,4-D. By keeping their supply of 2,4-D on barges, they are able to keep it close to the treatment area, which reduces the amount of time required for the empty helicopters to come back for re-loading.

Airboats are frequently used for aquatic herbicide application because they are able to travel through or even over the densest stands of aquatic weeds, which would be impassable to any other type of boat (Fig. 12.3). Since both the propeller and the rudder are in the air instead of the water, they do not become entangled in the weeds. The boats have flat bottoms which, when necessary, are capable of sliding up out of the water and gliding over a compressed mat of vegetation. Airboats are also very fast, which reduces the time required for application and, therefore, the cost.

Fig. 12.1.   Applying a liquid herbicide to a small pond with a garden-type sprayer. Note that the nozzle is beneath the surface to prevent drifting of the spray. *Courtesy of N.J. Coop. Extension Service.*

In large-scale operations where hand-held sprayers are not practical, liquid herbicides are usually dispensed with small pumps operated by gasoline engines. Instead of diluting the concentrated herbicide in advance, which would necessitate carrying a large quantity of diluted material in the spray boat, water is usually pumped out of the lake at the same time that concentrated herbicide is pumped from its container. The water and herbicide are then mixed in the proper proportions in a special mixing valve, and applied to the lake.

In the case of floating weeds such as water hyacinth, water lettuce, or emergent weeds such as cattail or spatterdock, the diluted herbicide is normally sprayed directly on the foliage. In the case of submersed weeds such as the pondweeds, elodea, coontail, and hydrilla, the diluted herbicide is normally injected beneath the surface of the water through a hose or a series of hoses trailing behind the boat. This has the advantage of placing the herbicide closer to the weeds to be controlled and allows greater flexibility in vertical distribution of the material. It also has the added safety advantage that the spray cannot be blown by the wind onto the applicator or onto surrounding shoreline areas.

Fig. 12.2. Applying granular herbicide to a small pond with a hand-held, rotary-type granular spreader. Note the respirator and goggles which prevent dust from getting into the applicator's eyes and respiratory system. *Courtesy of N.J. Coop. Extension Service.*

When applying herbicides to terrestrial areas, it is of utmost importance that the material be spread as uniformly as possible and that areas of overlap and skipped areas be held to an absolute minimum. In aquatic areas, perfectly even coverage is not as important. Because the herbicide will move short distances through the water by diffusion and by the action of currents and turbulence, it will have a tendency to become more evenly distributed after it has been applied. Small areas that were skipped in the initial coverage will be filled in from surrounding areas and herbicide will tend to move away from heavy concentrations where overlaps have occurred. Nevertheless, applications should be made as uniformly

Fig. 12.3.   An air boat being used to spray water hyacinth in Florida.

as possible. For large-scale operations, flags are sometimes put out on buoys or poles to serve as reference points to guide the movements of the spray boat or airplane. In one operation in the 1950s, the operator of the boat was equipped with a two-way radio, and his movements were guided by voice commands from a person on shore who sighted along the predetermined path of the boat with a transit. Such precise navigation has been found to be unnecessary and is rarely resorted to at the present time.

A regular pattern of application is usually followed to insure that no large areas will be missed. A zig-zag pattern such as the one shown in Fig. 12.4 is usually suggested. If time and cost are not severe restrictions, it is best to cover the area twice, using one-half of the herbicide each time. When the area is covered twice, the direction of movement the second time is usually at right angles to the original direction of movement as shown in Fig. 12.5.

When the biomass of weeds to be controlled is very large in relation to the volume of water, the danger of suffocating fish can be reduced by not treating the entire body of water at one time. It is usually recommended that one-half or one-third of the area be treated and that the weeds in this area be allowed to decompose before the next section is treated. This wiil give the fish an untreated area into which they can move in the event the dissolved oxygen becomes critically low in the treated area.

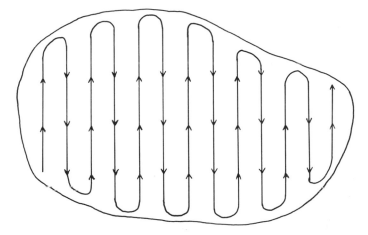

Fig. 12.4.   Herbicides should be applied in some sort of regular pattern to insure even coverage.

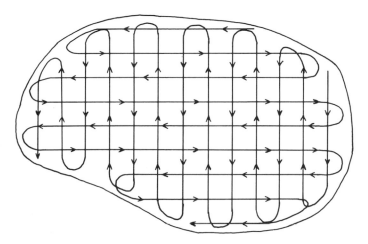

Fig. 12.5.   If double coverage is used when applying herbicides, the direction of travel for the second coverage should be at right angles to the first.

CALCULATION OF HERBICIDE RATES

The rate of application (the amount of herbicide to be used in a given situation) should never vary from the rates stated on the label. Using too little material may result in unsatisfactory control and using too much is not only illegal but may result in environmental damage as well. It is therefore extremely important that the proper amount of herbicide be used.

Lakes and Ponds

Some application rates are based on surface area of water to be treated and some are based on volume of water to be treated. This will be predetermined by the type of herbicide being used and will be clearly stated on the product label. In either case, the first step in calculating the amount of product to apply is to determine the surface area to be treated. The area of large lakes and reservoirs is almost always available from various county or state offices, but for smaller bodies of water it may have to be calculated.

This can sometimes be done by direct measurement with a surveyor's tape. For roughly rectangular, circular, or triangular ponds the necessary distances can be measured and the area of the corresponding geometric figure can be calculated. For irregularly shaped bodies of water, the surface can usually be divided into distinct areas each of which has a more or less regular shape with which you can deal. Figure 12.6 shows an irregularly shaped pond, which has been divided into a semicircle, a rectangle, and a triangle. The surface area of each can be calulated separately, and they can then be added together to obtain the area of the entire pond.

If the amount of herbicide to be used is based on the water volume, it is necessary to determine the average depth of the water as well as the surface area. Using half of the maximum depth is not sufficiently accurate. It is necessary to approximate the true average depth more accurately than that. If a contour map of the lake or pond is available, transects can be drawn on it and an average can be calculated of all those points at which a transect crosses a contour line. It is important to average in the zero depth values where the transects cross the pond's perimeter. If a contour map is not available, the only recourse is to take soundings from a boat and, in effect, make a contour map. In deep waters a sounding line is

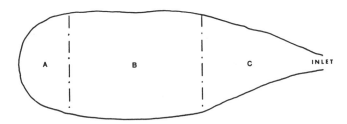

Fig. 12.6.   When calculating the surface area of an irregularly shaped pond, divide it into areas which are approximations of regular geometric figures. A is a semicircle, B is a rectangle, and C is a triangle.

necessary but in shallow ponds a surveyor's range pole makes an excellent sounding device since it is rigid and easy to handle and is already calibrated in 1-foot increments (Fig. 12.7). When making soundings, one cannot ignore the depths of zero at the water's edge or the calculated average depth will be too large.

Application Based on Surface Area.   When herbicides are applied on the basis of surface area, the amount to be used is almost always expressed in terms of pounds of herbicide per acre of water (lb/acre). The problem is that most commercial herbicides do not consist of 100% active ingredient. As explained earlier, they contain various extenders, binders, carriers, surfactants, and solvents in addition to the chemical having the

Fig. 12.7.   Using a surveyor's range pole to determine the average depth of a shallow body of water. *Courtesy of N.J. Coop. Extension Service.*

basic herbicidal properties. At one time the common practice was to make herbicide recommendations in terms of the amount of active ingredient to be applied per acre of water. In recent years the trend has been to make recommendations in terms of the amount of commercial product to be applied per acre of water. This is an important distinction. If a commercial granular herbicide contains 20% active ingredient by weight, and the proper rate of application is 100 lb of commercial product per acre, a calculation mistakenly based on 100 lb of active ingredient per acre would result in a 500% overdose.

In the case of liquid herbicides that are applied on the basis of surface acreage, recommendations are made in terms of pints or quarts of commercial product per acre or in terms of pounds of active ingredient per acre. In the latter case, calculations are not difficult because the product label will always give the number of pounds of active ingredient contained in one gallon of commercial product.

Application Based on Water Volume.   Application rates based on water volume are almost always expressed in terms of "parts per million." Unless specifically stated otherwise, this means "parts per million *by weight*," and it represents the weight of herbicide per unit weight of water to be treated. Thus, 1 gram of herbicide in 1 million grams of water equals 1 part per million (ppm). Similarly, 3 lb of herbicide in 3 million lb of water equals 1 ppm. It is the *ratio* of weights that determines ppm, not the absolute weight. Virtually all aquatic herbicide rates based on water volume instead of surface area are calculated on the basis of active ingredient, not commercial product.

It has been established that 1 acre-foot of water weighs 2.7 million lb. (An acre-foot is the amount of water required to cover 1 acre to a depth of 1 foot.) It follows, therefore, that dissolving 2.7 lb of any material in 1 acre-foot of water will result in a concentration of 1 ppm. Below are two examples of calculations of the amount of herbicide to be used in certain situations, based on water volume.

Example 1. The herbicide to be used is a powder that contains 80% active ingredient by weight. The pond to be treated contains 16.6 acre-feet of water. The label says to use a rate of 0.5 ppm active ingredient in order to control elodea. To calculate the total amount of commercial product to put in the pond:

a.   0.5 ppm × 2.7 lb = 1.35 lb active ingredient for each acre-foot
b.   1.35 lb × 16.6 acre-feet = 22.4 lb active ingredient for the entire pond
c.   22.4 lb/0.80 (80% active ingredient) = 28.0 lb of commercial product required for the pond.

Example 2. The herbicide to be used is a liquid containing 2 lb of active ingredient per gallon. The volume of water to be treated is 9.9 acre-feet and the concentration desired is 3 ppm. Calculations are made as follows:

    a.   3 ppm × 2.7 = 8.1 lb active ingredient needed for each acre-foot

    b.   8.1 lb/acre-foot × 9.9 acre-feet = 80.2 lb of active ingredient needed

    c.   80.2 lb/2 lb/gal = 40.1 gallons of commercial product needed

Flowing Water

Calculating herbicide application rates for streams, ditches, and other flowing waters is considerably more difficult than calculating rates for the relatively static waters of lakes and ponds. The fact that the current begins to move the herbicide as soon as it is applied makes the entire operation different. Since the herbicide is constantly moving, there is a new factor to be considered, and that is the concept of contact time. In static waters it is assumed that the herbicide will be in contact with the plant until it is absorbed, decomposed, or inactivated in some other way. In flowing waters the herbicide may only be in contact with the plant for a matter of minutes before being swept away by the current.

Since the amount of herbicide actually getting into the plant is a function of both concentration and time, both of these factors are manipulated simultaneously in the treatment of flowing waters. The herbicide is normally not spread over a wide area with a sprayer, but is introduced into the water at a single point, just upstream from the point where the weeds are to be controlled. One strategy is to introduce the herbicide into the water very slowly so that the concentration in the water is not high at any time but plants downstream will be subjected to a low concentration for a prolonged period. When using this system, herbicide is sometimes metered into the water continuously but very slowly over a period of several days.

The other strategy is to do the exact opposite of that which is described above. A very large amount of herbicide is introduced into the water all at once. This results in a "block" of heavily treated water moving downstream as a unit. As it passes an individual weed, that plant is exposed to a heavy concentration of herbicide for a short period of time. The former strategy results in chronic phytotoxicity, while the latter results in acute phytotoxicity. Both strategies are depicted graphically in Fig. 12.8.

Most chemical control of weeds in flowing waters has been done in irrigation canals, drainage ditches, and other man-made systems. Very little research or operational weed control has been done in natural rivers or streams. This is probably due to the fact that man-made systems are

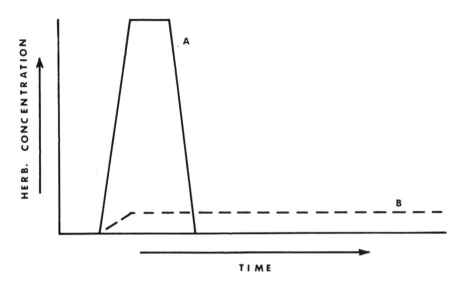

Fig. 12.8.   Herbicide concentrations in a flowing canal at a point downstream from the point of herbicide application, using two different treatment techniques. Curve A, A high concentration over a short period of time. Curve B, A low concentration over a long period of time.

relatively straight and uniform in cross section, which results in stable, uniform flow. Natural rivers and streams, on the other hand, have turns and bends, shallows and rapids, eddies and pools, all of which create turbulence and nonuniform flow. This lack of uniformity, coupled with the addition of fresh water from tributaries makes the calculation of herbicide rates almost impossible.

BIBLIOGRAPHY

MOORE, G.T., and KELLERMAN, K.F.   1904.   A Method of Destroying or Preventing the Growth of Algae and Certain Pathogenic Bacteria in Water Supplies. Bureau of Plant Industries Bull. 64, U.S. Dept. Agric., Washington, DC.
YOUNGER, R.R.   1965.   An aquatic application service. Hyacinth Control J. 4, 19-20.

13

Mechanical Control of Aquatic Vegetation

Mechanical weed control may be defined as the physical destruction of weeds or the physical removal of weeds from the site where they are causing a problem. Mechanical control is the most obvious form of weed control and the oldest form practiced by mankind. In terrestrial areas, mechanical weed control goes back to the time before written history. Every time a person pulls a weed from his garden, his path, or even from a flower pot, he is practicing mechanical weed control. In aquatic areas, mechanical weed control also predated all other methods.

HAND METHODS

The simplest, most primitive means of mechanical control are hand-pulling, hand-raking, and hand-cutting. Hand-pulling is slow, laborious, and expensive if paid labor is used. It is, nevertheless, still used in some parts of the world where labor is cheap. It is also useful in very small areas where emergent vegetation has just begun to establish itself. By pulling out a few plants when they first appear, a future problem of much larger proportions may be averted. Hand-raking is also slow and labor intensive. It is used in much the same types of situations as hand-pulling, but where the problem weeds are submersed instead of emergent species. Hand-raking might be used, for example, to keep a small bathing area free of weeds around a private dock.

Hand-cutting, while also slow and laborious, has some special applications that should probably be used more frequently. One very specific use is in the control of cattail. If mature cattail plants are cut off *beneath the surface of the water* they will not grow back. When cut above the surface of the water, considerable regrowth will usually occur. The subterranean portions of the plant are dependent on the aerial parts for oxygen and when this supply is cut off, the subterranean structures die before they can generate new shoots. If the cutting is done even slightly above the water surface, the oxygen supply is not cut off and the plants live and generate new shoots.

Spatterdock, in stands of limited size, can also be controlled by hand cutting. The petioles are very brittle and can be cut easily with a hand-held sickle or scythe. Unlike cattail, spatterdock will grow new leaves whether it is cut beneath the surface or above the surface. If the regrowth is cut regularly at intervals of two weeks or less, the rhizomes will begin to deplete their supply of stored food and will eventually die. In the Northeastern United States, it requires five or six cuttings at regular two-week intervals to kill the plant. On large stands, this is a very time-consuming practice, but in view of the fact that there are no effective biological controls, and since chemical controls are only marginally successful, it is sometimes worthwhile. Other floating-leaved or emergent species can doubtlessly be controlled in a similar way, but research is needed to determine the frequency of cutting and number of cuttings necessary to achieve control.

EARLY MECHANICAL DEVICES

Because of the obvious disadvantages inherent in hand methods of aquatic weed control, numerous mechanical devices have been employed over the years. The earliest were simple rakes or bars, weighted to keep them from riding up over the masses of weeds, and dragged behind a boat. Most were not made specifically for that purpose but were adapted pieces of discarded farm machinery. They all functioned by scratching or scraping the surface of the hydrosoil, thereby dislodging or breaking off the weeds. A similar stategy was used in irrigation canals in the western United States and was known as "chaining." A length of logging chain was fastened between two tractors, one on either side of the canal. The tractors dragged the chain along the canal bottom, dislodging submersed aquatic plants as they traveled along. Sometimes a tangled mass of barbed wire was fastened to the middle of the chain to add to its effectiveness.

Among the earliest devices manufactured specifically for aquatic weed control were cutter bars designed to be dragged along the bottom, cutting the weeds off at soil level. Some were V-shaped with teeth on the inner sides of the V, and some were simply straight, knifelike blades. They dulled easily and required relatively flat or gently undulating bottoms devoid of rocks, stumps, and other obstacles.

A more advanced system for cutting submersed weeds utilized reciprocating cutter bars similar to those used on tractors for mowing roadside grasses. The very earliest of these devices were strapped beneath an ordinary rowboat by means of canvas straps which came up over the gunwales of the boat and buckled on top. They were powered by a portable gasoline engine that was placed in the boat. A major disadvantage was that they could not be adjusted for depth of water once they were in operation.

Far more sophisticated equipment was devised for the special problem of controlling water hyacinth in Florida and other southern states. One such piece of equipment was called a saw boat. It consisted of a boat with numerous circular saw blades mounted approximately 1 in. apart on a horizontal shaft in front of the boat. In operation, the blades were lowered into the water to a depth of one-half their diameter and they revolved at 900–1000 rpm. As the boat progressed through the floating mass of water hyacinth, the revolving blades shredded the plants.

Another device of the same era was known as the "Kenney Crusher." This was a flat bargelike boat that lifted the mass of water hyacinths out of the water onto a conveyor belt that carried it back to one or more pairs of corrugated metal rollers that crushed the plants between them. The crushed plants were then carried to the back of the boat where they were returned to the water. Although the saw boats with their whirling blades and the crusher boats both had disadvantages that will be discussed below, they were, for many years, major weapons in the fight against water hyacinth.

Other types of mechanical equipment have been investigated but were used only experimentally or to a very limited degree on an operational basis. One of these was a hydraulic "churning" device that forced water through a nozzle at very high pressure. The nozzle was held near the bottom and directed downward so that the high-pressure stream churned up the substrate and dislodged the submerged vegetation growing there.

All of the early devices discussed above had at least one major drawback. The weeds that were cut off, uprooted, crushed, or shredded, remained in the water. If they were cut off, many of the cuttings floated free in the water until they touched down somewhere, rooted, and became new plants. Cuttings can live almost indefinitely in a free-floating condition, since they do not dessicate as plants would on land, and they are able to absorb dissolved nutrients through the leaf and stem surfaces. The basal portions of the original plants frequently resprouted, and two plants existed where only one had occurred before. Since the cuttings were free to move with currents, tides, or winds, weed infestations were frequently spread to new locations. Plants uprooted by rakes, chains, or hydraulic churning were also free to drift to new areas and re-establish themselves elsewhere.

Although some plants survived crushing or shredding, most were killed and were not able to continue growing. The problem with the crushers and the saw boats was that they returned to the water the mass of plant material, which immediately began to decompose. This organic decomposition depressed dissolved oxygen levels and sometimes suffocated fish. In addition, the inorganic nutrients contained in the plants were released back into the ecosystem where they were available to trigger algae blooms or stimulate the growth of more vascular plants. In order to overcome these problems, sophisticated harvesters have been developed that cut off submerged vegetation and then remove it from the water so that cuttings cannot re-root and dead plants will not decompose in the aquatic ecosystem.

MODERN HARVESTERS

Modern aquatic weed harvesters come in many sizes and with a wide variety of supplemental equipment, but they all work on the same basic design. The cutting boat itself is a flat, bargelike boat, which is self-propelled and capable of operating in very shallow water. It is typically equipped with paddle wheels instead of a propeller, which gives it greater maneuverability and shallower draft. Two arms extend from the front of the boat and a reciprocating mower bar is fastened between them in a horizontal position. The arms can be raised or lowered by means of a hydraulic system so that the mower bar can be adjusted to varying water depths or can be lifted out of the water entirely if desired. Some harvesters also have vertical mower bars extending upward from both ends of the horizontal bar (Fig. 13.1). The function of these vertical mower bars is to cut the mass of weeds to be harvested free from the weeds growing on either side. This is especially important when harvesting weeds that grow in thick, tangled mats extending from the bottom all the way up to the water surface. The weed-free path left behind such a harvester has a flat bottom and vertical sides.

Behind the mower bar (or bars) is a conveyor belt of some sort to pick up the weeds that have been cut off (Fig. 13.3). On some harvesters this is a simple, fixed-position conveyor, whose forward end is beneath the water, that picks up the cuttings as they float to the top. Other models are much more complex and have small accessory conveyors at the sides that feed the cut material toward the center of the swath where it is picked up by the main conveyor. In any case, the cut mass of weeds is removed from the water and lifted up onto the boat.

In the simplest and least efficient harvesters, the weeds are simply piled on the flat metal deck of the cutting boat. When a full load is collected, the boat proceeds to shore where the weeds are off-loaded by

Fig. 13.1.   A weed cutter with horizontal and vertical cutting bars, designed for mounting on an ordinary boat. *Courtesy of Air-Lec Industries, Inc.*

hand and loaded onto waiting dump trucks to be disposed of. This is highly inefficient because during the time the boat is carrying a load of weeds to shore and while it is returning to the harvesting site, it is not doing any harvesting. An expensive, specialized piece of equipment is being used as a transport barge. To circumvent this problem, some harvesting systems utilize auxilliary boats, which are much cheaper than the cutter boat and are used only to transport the harvested weeds to shore. In some systems the transport barges "lock" onto the rear of the cutter boat and the conveyors dump the weeds directly onto the deck of the barge. As soon as one barge is full, it is disconnected and heads for shore and another barge is hooked on to the cutter so it can resume harvesting immediately. In other systems, the weeds are temporarily stored in a large hopper on the rear of the cutter boat. When the hopper is

Fig. 13.2.   An underwater mower permanently mounted on a flat barge which will operate in 6 in. of water. *Courtesy of Hockney Co.*

full, a transport barge comes alongside and the hopper, which is hydrauli-cally operated, dumps its load onto the barge. In either case, the cutter boat is free to harvest plants most of the time and is interrupted only briefly while the barges are being changed or the hopper is being dumped.

The number of transport barges required depends on how far the harvesting site is from the shoreline location where the weeds are loaded onto trucks. It also depends on how long it takes to unload the barge after it reaches shore. In order to speed up the latter operation, shore convey-ors are available to lift the weeds from the barge onto the truck (Fig. 13.6). It still may be necessary to load the weeds onto the conveyor manually with pitch forks. This transfer of weeds from barge to truck is the least mechanized part of the whole operation and is labor intensive. The number of trucks and drivers required is dependent primarily on the distance from the shoreline loading site to the final disposal site.

Fig. 13.3.   A mower-harvester that picks up the cut weeds on a
conveyor belt and dumps them on the flat deck of the boat.
Courtesy of Rolba Co., Ltd.

Effectiveness of Harvesting

Modern aquatic weed harvesters do an effective job of cutting off the
tops of submerged plants and removing them from the water. They leave
a path in their wake that is almost entirely free of vegetation to the depth
of the cutter bar and practically no fragments remain in the water to
decompose or to float away and root at another location. Because of the
design of the boats, they are able to operate in very shallow water and
very few places in a lake or reservoir are inaccessible to them. Harvesters
are therefore capable of effecting good initial control and of doing it in an
environmentally sound way. The major problem with harvesting is that
the whole plant is not destroyed. The tops of the plants are removed but
the basal portions remain in place and are relatively unaffected. Since
submerged aquatic plants have excellent powers of regrowth, the basal
portions normally produce new shoots and in time the weed problem is
back. The important question is, "How fast will they grow back?" This

Fig. 13.4.   An underwater mower fitted with a weed rake, ready to drop a load of cut weeds onto the bank. *Courtesy of John Wilder (Engineering) Ltd.*

Fig. 13.5.   A large mower-harvester with several conveyer belts for automated handling of cut weeds. *Courtesy of Mud Cat Div., National Car Rental System.*

Fig. 13.6. A mower-harvester off-loading onto a shoreline conveyor which lifts the weeds directly into a waiting dump truck. *Courtesy of Mud Cat Div., National Car Rental System, Inc.*

depends on numerous factors such as the species of plant, the time of year, the nutrient content of the water and substrate, and probably other factors of which we are not even aware. Species with underground food-storage organs such as tubers and rhizomes and species with perennating organs such as turions tend to grow back more quickly and more densely than species that do not have these organs. In cooler temperate climates, with short growing seasons, a single harvesting operation may result in acceptable weed control for an entire season. In warmer temperate climates several harvests may be necessary, and in subtropical and tropical climates, frequent harvesting may be necessary throughout the year.

Nutrient Removal

One of the major arguments in favor of harvesting aquatic weeds instead of controlling them by chemical means is that nutrient elements contained in the plants are removed from the ecosystem. Several interesting studies have been conducted to determine how effective harvesting is in removing nutrients and how the amount of nutrients removed compares to the amount that accumulates in a given lake each year. A

TABLE 13.1. Morphometric Parameters of Lake Sallie and Chemung Lake

Parameters	Lake Sallie	Chemung Lake
Area	530 ha	858 ha
Mean Depth	6.35 m	1.80 m
Max. Depth	16.5 m	6.10 m

Source: Neel *et al.* (1973); Wile (1978).

brief summary of two such studies is given below. Phosphorus is used as an example because it was thoroughly studied in each case and because it is the nutrient most often implicated as a causal agent in aquatic weed problems.

The first study was sponsored by the United States Environmental Protection Agency and was conducted in Lake Sallie, Minnesota (Neel *et al.* 1973). The second was conducted by the Ontario Ministry of the Environment in Southern Chemung Lake, Ontario (Wile 1978). Some morphometric parameters of the two lakes are presented in Table 13.1. It is obvious from this table that the lake basins are very different in their basic conformations. Southern Lake Chemung is considerably larger in surface area but has a mean depth less than one-third that of Lake Sallie. This is an important factor that will be discussed later.

Table 13.2 presents some basic facts on the phosphorus budgets of both lakes. It can be seen that the annual phosphorus input to Lake Sallie far exceeds the annual input to Southern Lake Chemung. The annual loading rate (the amount of phosphorus retained in the lake each year) is also much greater in Lake Sallie. Even so, harvesting operations in Lake Sallie for one year (1970) removed only 100 kg (220 lb) or 1.3% of the annual loading rate. In Lake Chemung, on the other hand, harvesting operations for one year (1975) removed 560 kg (1235 lb) or 92% of the annual loading rate. How can such a large difference be explained?

TABLE 13.2 Annual Phosphorus Budgets of Lake Sallie and Chemung Lake

	Lake Sallie	Chemung Lake
Total P in inflow	15646 kg	1190 kg
Total P in outflow	7759 kg	580 kg
Retained in lake	7887 kg	610 kg
Removed in harvested weeds	100 kg	560 kg
% of net annual retention removed in weeds	1.3%	92%

Source: Neel *et al.* (1973); Wile (1978).

While there were many variables in these two studies that could have contributed to the difference in effective phosphorus reduction, two factors stand out as being the probable major causes. First is the enormous difference in the amount of phosphorus entering the two lakes each year. In Lake Sallie the submerged vegetation was just not able to absorb a significant fraction of such a large amount of the nutrient. Even with luxury consumption, there is a limit to the percentage of phosphorus plant tissue can contain. The second factor is the shape of the two lake basins. Southern Lake Chemung, being relatively shallow, was inhabited by submerged vegetation over approximately 50% of its area, while Lake Sallie, being relatively deep, was inhabited by submerged vegetation over only 34% of its surface area. Thus, there were relatively fewer plants in Lake Sallie to absorb phosphorus and to be harvested.

Other, roughly similar, studies have been conducted in other localities, and most of their results fall somewhere between the extremes of Southern Chemung Lake and Lake Sallie. Harvesting submerged aquatic weeds obviously removes plant nutrients from the system, but the relative importance of such removal in comparison to the rate of accumulation, varies greatly, and it is impossible to make broad generalizations regarding the overall value of harvesting in slowing down or reversing eutrophication. Even where harvesting equals or surpasses the annual loading rate, release of phosphorus from the sediments may continue for a number of years, resulting in little change in ambient dissolved phosphorus in the water.

Cost of Harvesting

The cost of harvesting aquatic weeds has been frequently cited as one of its primary disadvantages. Certainly the initial cost is high due to the capital investment required to purchase the harvester and the auxilliary equipment. When this cost is amortized over a period of years, however, it becomes much more reasonable. The cost of mechanical harvesting has been estimated by various workers under a wide variety of conditions and with different types of harvesters. Among the variables affecting operating costs are cost of local labor, weed density, distance from harvesting site to shoreline, number and type of transport barges required, and number of dump trucks and drivers required. Another important, but frequently overlooked, factor is the percentage of time that the harvester has to spend in nonharvesting activities such as being repaired and moving from one location to another.

With so many variables, it is not surprising that harvesting costs have been estimated as low as $60.00/acre/harvest and as high as $600.00/acre/harvest (Smith 1979). Even in cases where the cost per acre is fairly low, final disposal of the weeds is a problem and adds considerably to the overall expense. One solution to this problem would be to find an economic use for the harvested weeds so they could be sold. Even if they were not sold at a profit, the income would help to alleviate the cost of harvesting. Although many uses have been proposed, none has proved to be very practical, and most harvested weeds are still disposed of in a landfill. A more detailed discussion of uses of aquatic plants can be found in Chapter 14.

DREDGING AND BULLDOZING

Dredging a body of water or draining it and removing accumulated sediments with a bulldozer, drag line, or other earth-moving equipment are highly effective means of controlling aquatic weeds. Not only are the tops of the plants removed, as in harvesting, but roots, rhizomes, seeds, and other plant parts on or in the substrate are also removed. Furthermore, large amounts of phosphorus and other plant nutrients are removed with the sediment. In addition, by deepening the water the amount of light penetrating to the bottom is reduced. If most of the bottom area can be excavated to below the compensation depth, weeds will not become a major problem. In any dredging, bulldozing, or other excavating procedure, the sides of the pond or lake should be made to slope steeply to deep water and the bottom should be as deep as is reasonably practical in order to keep weed problems to a minimum.

The major disadvantage of this type of weed control is the cost. Dredging is very expensive and the use of a bulldozer or drag line on all but the smallest ponds is also costly. As with harvesting, sale of the dredged sediments would help to alleviate the cost of dredging, even if a net profit was not realized. Unfortunately, the sediments are not as useful as one might expect them to be. They are composed of very fine particles and have very poor "structure." This means the particles are not aggregated together into larger particles. Because of this, the sediments tend to become very hard and crack when they dry out, and they become very soft and jellylike when they become wet. As a result they do not make good top soil. They do find certain specialty uses, however. For example, they can be mixed with other materials such as peat moss, sand, and vermiculite to make potting mixes for use in the horticultural industry. When the dredged sediments cannot be sold, disposal is often a troublesome and costly problem.

SHADING

Attempts to control aquatic weeds by reducing their light supply have been successful in some instances. The use of fish that keep the water turbid and the use of phytoplankton blooms have been mentioned in Chapter 11. Other methods that do not utilize living organisms are discussed below.

Plastic Film

Sheets of black plastic, weighted and spread over weed beds have been employed to block out the sun. Although partially successful, the plastic sheets are very difficult to put in place. The biggest problem, however, is that gases produced by decaying organic matter in the substrate and later by the decomposing weeds, accumulate in huge bubbles under the plastic. The buoyancy of these gases displaces the plastic even when heavily weighted. Cutting parallel slits in the plastic allows the gas to escape but weakens the material and causes it to tear more easily when being handled and put in place.

Fiberglass Mesh

A recent development that overcomes most of the disadvantages of black plastic film is vinyl-coated fiberglass screening. This material is manufactured and sold specifically for this purpose. The openings in the mesh are 1 mm, which is sufficient to allow gas to escape but apparently excludes enough light to cause the covered plants to slowly die. Some workers (Perkins *et al.* 1979) feel that control may be a result of physical contact and crowding as much as it is a result of light reduction.

The mesh is expensive, but it is reuseable and can be moved from one location to another as the weeds die. Since it does not trap air, and has a density greater than that of water, it is easier to put in place than plastic film.

Dyes

The use of dyes to absorb sunlight and prevent its penetration into the water is, in a sense, a form of chemical control. However, since the action is not herbicidal in the usual sense, it is being discussed here, along with mechanical methods. One type of dye is being sold at the present time for the specific purpose of shading submerged aquatic weeds. It is a non-toxic, water-soluble compound that turns the water bright blue. In addition to reducing light intensity, the dye drastically alters the spectral composition of the light reaching the plants. Both factors probably play a role in reducing their growth.

The use of dyes is most effective early in the growing season when the plants are still near the bottom. After the plants reach the surface, it is recommended that they be removed by harvesting or with herbicides, followed by the dye to suppress regrowth. The dye is very persistent and will last for long periods of time if it is not flushed out of the pond. It is most useful in small bodies of water that are completely static. Even a slow rate of water exchange will soon dilute the dye and destroy its effectiveness. For the same reason it cannot be used to treat a small area in a large body of water. It is quickly diluted by water from surrounding, untreated areas and its effectiveness is lost.

DRAINING AND FREEZING

Draining the water from an area infested with aquatic weeds has long been practiced as a control measure. This may involve complete dewatering of a pond or lake, or it may involve lowering the water to a predetermined level to expose only shallow areas where weeds are known to occur. Where draining or lowering of water level is physically possible and where it does not interfere with other water uses, it has been a reasonably successful method of weed control. Careful thought must be given to the consequences of draining a body of water. The obvious questions to consider are things such as the effects on the fish population and temporary loss of water for irrigation, stock watering, and recreation. Other unforeseen problems may develop, however. In one instance, the late fall drawdown of a 300 acre lake caused shallow wells in the surrounding area to go dry. Naturally, the project had to be abandoned. In another instance, a farmer drained his farm pond for weed control purposes and his insurance company insisted that he refill it because it was the only source of water in the vicinity of his house and barn that was available for fire-fighting purposes.

The time of year at which the drawdown occurs is very important in determining the success or failure of the procedure. Virtually all researchers agree that the drawdown is most effective if done during the winter months. In colder climates a winter drawdown has the added advantage of allowing the exposed bottom to freeze, which increases the degree of weed control obtained. It has been shown that the freezing of exposed Eurasian water milfoil plants is more effective than freezing them in very shallow water (Stanley 1976). Frost penetration of at least 10 cm (4 in.) for no less than 3 weeks usually affords excellent control.

Even in climates where freezing does not occur or where frost penetration is not very deep, winter drawdown is more effective than dewatering at other seasons. It has been successful in Florida (Hestand and Carter 1974; Tarver 1980) and Louisiana (Richardson 1975; Manning and Johnson 1975). Officials of The Tennessee Valley Authority have used a combination of winter drawdown and herbicides for many years to control Eurasian water milfoil. During the winter months the water level is lowered to expose shallow areas and control the water milfoil growing there. During the summer months 2,4-D is applied to the deeper areas where the weed was not exposed during the drawdown period. They feel that dewatering is the most successful method available for water milfoil control (Smith 1971).

It is generally necessary to keep an area dewatered for a period of 2−6 months in order to effect good weed control. The duration of dewatering varies with the species of weeds involved, how well the substrate drains, and climatic conditions, including the amount of rainfall which occurs.

BIBLIOGRAPHY

HESTAND, R.S., and CARTER, C.C. 1974. The effects of a winter drawdown on aquatic vegetation in a shallow water reservoir. Hyacinth Control J. **12**, 9−12.

MANNING, J.H., and JOHNSON, R.E. 1975. Water level fluctuation and herbicide application: an integrated control method for hydrilla in a Louisiana reservoir. Hyacinth Control J. **13**, 11−17.

NEEL, J.K., SPENCER, A.P., and WINTFRED, L.S. 1973. Weed harvest and lake nutrient dynamics. Ecological Reseach Series, EPA-660/3−73−001. U. S. Environmental Protection Agency. Washington, D.C.

PERKINS, M.A., BOSTON, H.L., and CURREN, E.F. 1979. Aquascreen: a bottom covering option for aquatic plant management. *In* Aquatic Plants, Lake Management, and Ecosystem Consequences of Lake Harvesting. J.E. Breck, R.T. Prentki, and O.L. Loucks. (Editors). Inst. for Environmental Studies, Univ. of Wisconsin, Madison, WI.

RICHARDSON, L.V. 1975. Water level manipulation: a tool for aquatic weed control. Hyacinth Control J. **13**, 8−11.

SMITH, G.E. 1971. Resume of studies and control of Eurasian water milfoil (*Myriophyllum spicatum*) L. in the Tennessee Valley from 1960 through 1969. Hyacinth Control J. **9**, 23−25.

SMITH, G.N. 1979. Recent case studies of macrophyte harvesting costs: options by which to lower costs. *In* Aquatic Plants, Lake Management, and Ecosystem Consequences of Lake Harvesting. J.E. Breck, R.T. Prentki and O.L. Loucks. (Editors). Inst. for Environmental Studies, Univ. of Wisconsin, Madison, WI.

STANLEY, R.A. 1976. Response of Eurasian water milfoil to subfreezing temperature. J. Aquatic Plant Management **14**, 36−39.

TARVER, D.P. 1980. Water fluctuation and the aquatic flora of Lake Miccosukee. J. Aquatic Plant Management **18**, 19−23.

WILE, I. 1978. Environmental effects of mechanical harvesting. J. Aquatic Plant Management **16**, 14−20.

14

Utilization of Aquatic Plants

This chapter deals with the direct, purposeful utilization of aquatic plants by human beings. It does not deal with benefits which accrue to mankind indirectly such as the food value that aquatic vegetation provides to wildlife. The first part of the chapter covers use patterns in which plants are purposely planted, cultivated, and/or harvested for some particular use. The second part of the chapter covers use patterns in which plants are harvested in order to solve a weed problem and finding a use for them is a secondary consideration.

PLANTS RAISED OR HARVESTED FOR USE

Human Food

With the exception of rice, which is an agronomic crop and beyond the scope of this book, aquatic plants are not widely utilized as food by people. They were probably eaten more frequently in antiquity than they are today and more by primitive peoples than by more advanced civilizations.

In ancient Egypt water lilies were regularly harvested and even cultivated for human consumption. Herodatus, the Greek historian described the practice in the fifth century BC. Dioscorides and Pliny The Elder also wrote about it in the first century AD. The plants were gathered and dried and the seeds were then pounded or ground into flour, which was used to make bread. Other parts of the plants, including the carpels, were eaten raw. Some of the seeds were saved each year and were individually rolled in little balls of clay that were broadcast over the surface of the water to insure a crop for the following year (Sculthorpe 1967). It is interesting to note that in the early twentieth century a recommended method of establishing water lilies was still to roll the seeds in little balls of clay and to broadcast them over the surface of the water. Various species of water lilies and lotus are still cultivated in the Orient for their fruits, seeds, and rhizomes.

Water chestnuts are also widely cultivated in the Orient. The Chinese water chestnut of commerce is the corm of an emergent rushlike plant (Eleocharis dulcis), which is raised in flooded fields, often in rotation with rice. Additional water chestnuts are grown in the Mediterranean Region and shipped throughout Europe and North America. This plant should not be confused with an unrelated plant (Trapa natans), which unfortunately is also called water chestnut.

T. natans grows in the form of a floating rosette that is attached to the bottom by a long thin stem. Its curious, spiny fruits contain large fleshy seeds that are edible and are also cultivated and distributed commercially. In the 1960s this plant became established in the Mohawk River in New York State and in Chesapeake Bay. It became a serious weed problem and was eventually brought under control by means of an intense program of chemical and mechanical methods, but it remains troublesome in local areas in New York State.

In Africa, various tribes dig and eat the starch-laden rhizomes of wild water lilies. It is interesting to note that other tribes living nearby refuse to eat water lily rhizomes, even in times of famine, because they feel it is degrading to eat the same food as their neighbors whom they hold in contempt. In South Africa the flowers of an aquatic plant known as "waterblommetjie" have long been gathered and eaten by the local people in Cape Province (Goosen 1981). In the late 1970s the plant became extremely popular throughout the entire country as a result of a song recorded by a popular singer. Demand was so great that the plant all but disappeared from its natural habitat. Since 1980 waterblommetjie have been grown commercially, and research is underway at the University of Stellenbosch on cultivation methods.

Wild rice is the only aquatic plant whose seeds are regularly gathered and eaten in the United States. It is an annual grass, not closely related to cultivated rice. It is somewhat of a gourmet item, commanding a high price on the retail market. Although there have been numerous attempts to cultivate it, most wild rice is still gathered from naturally occurring stands in Minnesota and Wisconsin. The failure to grow wild rice under modern systems of cultivation is due primarily to the uneven ripening habit of the grain and the fact that kernels drop off the plant almost immediately upon ripening. The early dropping of ripened grain is referred to as shattering. Because wild rice ripens unevenly and because it shatters so badly, it must be harvested several times in order to collect most of the crop. Modern mechanized agriculture does not lend itself to multiple harvesting of grain crops. A breeding program is underway at the University of Minnesota to develop varieties that do not shatter so badly. Another problem with the cultivation of wild rice is that the seed must be kept cold and wet in order to keep it viable. Seed storage is thus expensive and troublesome. Being an annual, it must be replanted each year from seed.

At the present time, most wild rice is harvested by Indians who paddle canoes through natural stands, bend the stalks gently over the canoe, and tap the heads with a stick to dislodge the ripe grains. They may come back several times to the same stand as more seeds ripen.

The fresh foliage of several types of aquatic plants is eaten in various parts of the world. In the United States the only aquatic foliage eaten on a regular basis is watercress, which is used in salads and as a garnish. It is raised in shallow "beds" having a constant supply of cool, clear, flowing water, usually from underground springs. It is originally from Europe, but is now naturalized throughout the United States, including Hawaii, where there are numerous small watercress farms. Wild populations of watercress are also harvested in many parts of the world.

Biological Filtration of Water

The second reason for growing aquatic plants is for use in removing nutrients from either raw sewage (rarely) or, more frequently, from the effluent of sewage treatment plants. The effluent from sewage treatment plants, even efficient modern plants, normally has concentrations of phosphorus, nitrogen, and other plant nutrients that exceed those of natural waters in the area. It is desirable and in some cases mandatory that these nutrients be removed prior to the return of the water to the environment. Chemical removal is costly, and the removal of each major nutrient usually requires a separate operation.

Aquatic plants are being used in some localities to reduce the concentrations of these nutrients prior to the release of the water. Plants have the advantage that they absorb not only the essential nutrients, but other elements and even organic pollutants such as phenols (Wolverton and McKown 1976). They are also energy efficient because they utilize the energy of the sun. The plants are harvested periodically, which removes the absorbed substances, including nutrient elements, from the system.

The choice of plants is frequently determined by the ease with which they can be harvested rather than by their ability to absorb nutrients or their productivity potential. Free-floating plants and emergent plants are most easily harvested and therefore are the types utilized almost exclusively for this purpose. Fortunately, there are both free-floating and emergent species that have great absorptive capacities and are capable of growing very rapidly and producing large amounts of biomass in relatively short periods of time.

Emergent Species.   Emergent species have been used in Europe for a number of years as a final cleanup for sewage effluents. Various species of bulrush and giant reed are planted in shallow, artificial marshes, and the effluent is allowed to enter the upper end very slowly so that it is retained in the marsh for several days, at least, before flowing out at the other end. As it passes through, the roots of the plants absorb not only essential nutrients, but other elements as well. In some systems, the effluent flows through a series of such marshes, each with a different species of plant, before being returned to a natural watercourse. Some of the marshes have a more or less natural substrate of soil and some are composed of artificial beds of gravel, which are lined below with heavy plastic to prevent any percolation of the wastewater into the ground.

The marshes are drained periodically and the reeds or rushes are harvested with mechanical equipment that is often modified to allow it to operate on a soft, wet surface. The roots and basal portions of the plants are unaffected and they re-sprout, making it unneccesary to replant the marsh each time. In many parts of Europe the harvested plants are sold and processed for their fiber content.

In the United States the use of emergent aquatic plants in artificial marshes for removing nutrients from wastewater is in its infancy. Most such marshes are experimental rather than operational and with a few exceptions, they are in the southern states where they are functional throughout the year. A major problem is disposal of the plant material after it has been harvested. There are no processing facilities for removing the fiber and the supply at this time is too small to make such a facility worthwhile. Other possible uses are being investigated and will be discussed later in this chapter.

Free-Floating Species.   Free-floating species have the tremendous advantage that they can be harvested by merely skimming them off of the water's surface. Partial harvests can be made on a periodic basis without disrupting the system in any way. The two groups of floating plants most often used are water hyacinth and the duckweeds. Both are vigorus, competitive growers that can attain enormous biomass in short periods of time through vegetative reproduction.

In tropical or semi-tropical climates, water hyacinth is usually the plant of choice. The very characteristics that make it one of the world's worst weed also make it one of the most useful plants for wastewater treatment. Its rapid growth and high biomass production mean that it is able to absorb large amounts of nutrients in a short period of time. Its dense pattern of growth prevents sunlight from penetrating into the water. This, along with the organic matter in the wastewater and that contributed by the water hyacinth, maintains an anaerobic condition beneath the

vegetative mat. Under these conditions denitrification occurs and additional nitrogen is removed from the water directly to the atmosphere as
elemental nitrogen. Because of other problems, such as odors, that are
associated with anaerobic conditions, many people feel that the water
should be maintained in an aerobic state, despite the benefits of denitrification.

Under ideal conditions, 20–40 tons of water hyacinth (wet weight) could
be harvested from 1 ha of water every day without reducing the population. Each day this would remove an amount of nitrogen equivalent to
the daily nitrogenous wastes of over 2000 people and an amount of
phosphorus equivalent to the daily phosphorus wastes of over 800 people
(National Academy of Sciences 1976). In the United States, disposal of the
harvested plants on a large scale remains a problem. In temperate climates, an alternative system would have to be used during the colder
months of the year when water hyacinths are unable to grow or else the
entire pond or lagoon would have to be contained in a large greenhouse.

Duckweeds, usually in mixed cultures of several genera, have also been
used as biological filters for wastewater. They, too, have tremendous
reproductive potential and are even easier to harvest than water hyacinth. Furthermore, they will tolerate much cooler temperatures, which
makes them suitable for use over a wider geographic area than water
hyacinth.

Some species of duckweeds can double their numbers in less than 3
days under ideal conditions (National Academy of Sciences 1976), and
laboratory tests have shown that more than 33,625 kg/ha (30,000 lb/acre)
can be grown each year (Truax et al. 1972).

Because various genera of duckweeds have been used and because
their growth rates and chemical compositions vary widely with the chemical compositions of the waters on which they are grown, it is difficult to
draw any general conclusions regarding the amount of nutrients duckweeds can remove. Table 14.1 gives the amount of dry plant tissue
harvested per month under one particular set of conditions and shows
the amounts of crude protein, nitrogen, phosphorus, and potassium
contained in that tissue. It should be understood, however, that under

TABLE 14.1 Rate of Removal of Nutrients from
Wastewater by Duckweeds

	Amount removed (kg/month/ha)
Duckweed (dry wt)	2800
Crude Protein	1120
Total N	185
Total P	60
Total K	65

Source: Culley and Epps (1973).

other conditions of water nutrient content, climate, and duckweed species composition, the figures might be considerably different.

Duckweed has one advantage not shared by water hyacinth or other species currently being suggested for wastewater cleanup. Because the individual plants are so small, they cover the water very efficiently and mosquito larvae are apparently unable to breathe at the surface. As a result, a heavy cover of duckweeds actually depresses mosquito populations instead of increasing them as many types of aquatic plants do (Culley and Epps 1973).

Fiber and Structural Material

Emergent aquatic plants have been used as sources of fiber and structural material since ancient times. The Egyptians used papyrus as a source of pulp for paper, and other peoples used similar plants tied in bundles for making boats. Reed boats made in this way have crossed the Atlantic Ocean and are still being used by Indians on lakes in South America.

Giant reed, which has an extensive range throughout the temperate and tropical parts of the world, has been utilized by different civilizations for such diverse purposes as thatching roofs, building fences, making musical instruments, and fashioning shafts for darts and small arrows. In modern times it has been harvested mechanically in Europe as a construction material.

The most consistent and largest scale use of giant reed has been in Romania where the reed swamps of the Danube delta have been very successfully exploited. The swamps are managed for giant reed production by manipulating water levels, and the plants are harvested with equipment designed and built specifically for that purpose. There are specific machines for harvesting in shallow water, on soft saturated soil, and on dryer, harder soil. The industry has developed and grown steadily since 1956 when the harvest was measured in tens of thousands of tons to recent years when the harvest has approached 200,000 tons (Boyd 1974).

Pulp mills process the reeds and extract the fiber, which is made into paper, cardboard, cellophane, and similar products. Pulp mill wastes and canes which have not been pulped are made into coarser products such as insulation and fiberboard, and they are even compressed into building blocks. Other products made from giant reed in Romania are alcohol, furfurol, and, reportedly, some fertilizer.

Other Uses

In some areas, particularly less developed countries, aquatic plants are gathered and used as food for livestock and/or as soil amendments. Since

Fig. 14.1.   *Phragmites australis* grows in very dense stands and attains considerable height. Here it is growing as a semi-aquatic plant, on a wet site but without free-standing water.

these uses are covered in the following section of this chapter, they will not be discussed further here.

UTILIZING HARVESTED WEEDS

In the preceding section of this chapter, we considered plants that were intentionally grown or intentionally harvested for use. In this section we will consider plants that are harvested primarily because they are causing problems and must be disposed of once they have been harvested. An economically feasible use for these plants would solve the problem of disposal and perhaps even help to alleviate the cost of harvesting.

The two most frequently proposed uses for harvested aquatic weeds are livestock feed and fertilizer. To put either of these on a practical, paying, commercial basis would require a relatively constant supply of weeds of uniform mechanical properties and chemical composition. In temperate parts of the United States where submerged weeds are harvested for only a few months each year, the supply is so irregular as to make the construction of processing plants impractical. Furthermore, harvested weeds from temperate areas tend to be comprised of multiple species, which differ from season to season and from lake to lake. This results in a lack of physical and chemical uniformity and makes use as livestock feed or fertilizer difficult.

Even in more southerly areas where weeds can be harvested through much of the year and where there are large stands of single species, the nature of the weeds varies with the stage of growth of the plants and with the chemical composition of the water in which they grew. Weeds from eutrophic waters have higher concentrations of organic and inorganic nutrients in them than those that come from waters of lower fertility.

Another problem is the high percentage of water in freshly harvested weeds. Even after so-called free water or gravitational water is drained off, freshly harvested aquatic weeds typically contain 90–95% water by weight. The composition of a mass of harvested Eurasian water milfoil is shown in Fig. 14.2. The high percentage of water makes transportation of the weeds very expensive. It is necessary to transport at least 90 tons of water to a processing plant in order to get just 10 tons of useful material. Processing plants would therefore have to be close to the sources of weeds in order to be practical.

Attempts have been made to express excess water at the harvesting site by means of portable presses, prior to transportation. All such trials have been unsuccessful because the pressure necessary to express significant quantities of water also ruptures many cell walls and the contents of the cells are lost. The remaining material consists mostly of cell walls and has a much higher ratio of cellulose to protein than the unpressed weeds, which makes it less valuable as a livestock food or fertilizer. In addition, if the pressing is done on the harvesting boat or directly at lakeside before loading the weeds onto trucks, the nutrients lost from the ruptured cells flow directly back into the lake and partially negate one of the benefits of harvesting, which is removal of nutrients from the aquatic ecosystem. Other methods of drying at the harvesting site, such as the use of solar energy, have all proved to be impractical.

Animal Feed

Fresh, green aquatic weeds are very poor animal feed because animals refuse to eat most species, and the high water content makes the nutri-

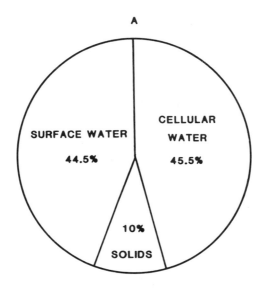

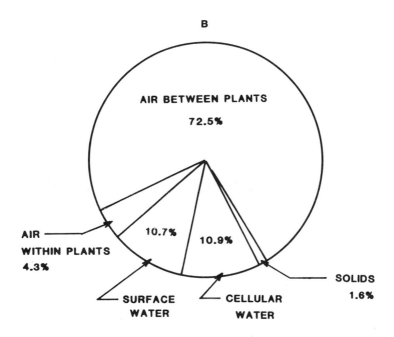

Fig. 14.2. Composition of a typical mass of harvested Eurasian water milfoil. (A) by weight, (B) by volume. *Redrawn from Koegel et al. (1973).*

tional value very low on a unit weight basis. Figure 14.3 compares the protein content of young alfalfa hay with that of eight species of aquatic plants on a dry weight basis and on a wet weight basis. On a wet weight basis, most of the aquatic species compare poorly with alfalfa but on a dry weight basis, five of the eight species exceed alfalfa in protein content. In order to obtain good quality feed, aquatic plants almost always have to be dried or processed in some way.

To assess properly the value of any material as a livestock food, one must consider its nutritive content, its palatability, its digestability, and its cost. No matter how nutritious a substance may be, it is of little use if the animals will not eat it or if they cannot digest it.

Nutritive Content.   The nutritive content of harvested aquatic weeds is highly variable. It varies with the many factors described above and,

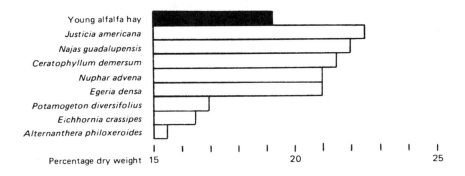

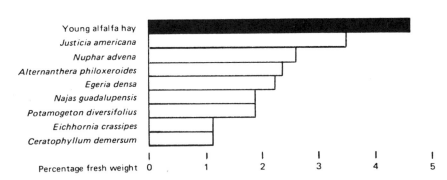

Fig. 14.3.   A comparison of the crude protein contents of young alfalfa hay with that of eight species of aquatic plants on a wet weight and a dry weight basis. *From Boyd (1974).*

unfortunately, most weed harvesting is not planned to coincide with the time of year or stage of growth when the weeds are most nutritious. Harvesting is undertaken when it is most convenient for the operators or when it interferes as little as possible with other water uses. As a result, the nutritional value of aquatic weeds is difficult to predict and generalizations are dangerous. As a very general rule, however, submerged aquatic weeds, harvested at or near their peak of nutritional quality, compare favorably with good quality hay. The duckweeds contain a much higher percentage of protein, and they will be discussed separately in a later section.

Palatability. The palatability of aquatic weeds, whether fresh or processed, is extremely poor when compared to normal feeds and forages. Almost all livestock will refuse to eat feed prepared entirely from aquatic weeds. Animals can be induced to eat feeds prepared from other materials with small amounts of dried, processed aquatic weeds mixed in, but the total food intake per animal is usually reduced and the animals either lose weight or, if they gain weight, the rate of gain is lower than for those animals not having aquatic weeds in their rations.

Some people have suggested that the poor palatability may be due to the odor of the weeds and others have suggested it may be due to the taste imparted by the unusually high percentage of tannins in many species. Whatever the cause, the poor acceptance of the plants by livestock is one of the major drawbacks of its use as a feed. For example, Ayoade *et al.* (1982) tried to feed water lettuce to a wide variety of domestic animals and albino rats. The water lettuce was offered in a finely chopped condition, both fresh and dry, and mixed in various proportions with grain. Five breeds of adult cattle would not eat the water lettuce and even tried to separate it from the grain when it constituted only 5% of the mixture. Sheep, goats, and horses rejected it completely in all forms and mixtures as did albino rats, even after they had been starved for 48 hours. Hogs, however, accepted the plants and grew reasonably well on them, although hogs have rejected other species of aquatic plants in other tests. Attempts to feed harvested weeds to fish have produced similar results. Liang and Lovell (1971) found that dried water hyacinth meal is not palatable to channel catfish *(Ictalurus punctatus)*. They will not eat food containing 40% or more of dried water hyacinth but can be conditioned, over a period of time, to eat foods containing lower percentages.

Digestability. Poor digestability is another problem that may be encountered when feeding aquatic plants to livestock. The digestability of proteins in feed is reduced by the presence of high concentrations of tannins. Tannin concentrations of 10% or more (on a dry weight basis) have been reported for a number of common aquatic species, including parrot feather, fanwort, watershield, and fragrant water lily. According to Boyd (1974), digestability of proteins in these plants would be greatly impaired.

Aquatic plants may also contain extremely high concentrations of minerals, which poses another problem. The ash content typically ranges from 10 to 30%, but values as high as 60% have been reported, and where carbonates have heavily encrusted the stem and leaf surfaces, the ash content could be even higher. As with all chemical components of aquatic plants, these values are highly site- and season-dependent as well as species-dependent.

Most estimates of digestability have been inferred from laboratory analyses of plants. Very few animal studies have been conducted to determine the percentages of nutrients that are actually digested and absorbed. Studies of this nature are badly needed with both ruminant and nonruminant animals in order to better assess the value of aquatic plants as livestock feed.

Cost.   Next to palatability, the cost of preparing animal feeds from harvested aquatic weeds is the greatest deterent to widespread implementation of this practice. Several factors contribute to the high cost. One of them is transportation. With a few exceptions, aquatic weed harvesting is conducted over widely scattered areas with no great concentration of weeds available at any one location. This necessitates inefficient collection and long distance hauling to get the weeds to the processing plants. The high water content of freshly harvested plants adds to the ineffiency and the cost of transportation.

Fresh aquatic weeds, particularly submerged species, decay rapidly, and they must be dehydrated quickly upon reaching the processing plant. As pointed out earlier, pressing the water out by force ruptures many cell walls and results in the loss of the cell contents and up to 50% of the crude protein. Drying in the conventional manner by means of hot air is extremely energy-intensive and therefore expensive. Processed feed prepared from aquatic plants in this manner cannot compete in cost with feeds prepared from terrestrial grasses and forages that contain less moisture to begin with and therefore require less drying.

Silage

Attempts have been made to ensile several species of aquatic plants with varying degrees of success. The major advantage of this method of preservation is that it is not necessary to dry the plants to the same moisture level as is necessary when they are to be processed into dry food. In the case of water hyacinth, for example, only about 50% of the moisture has to be removed in order to prevent putrefaction in the silo. It is also necessary to add an additional source of free carbohydrate, such as cracked corn, in order to obtain the fermentation required for proper preservation (Bagnall *et al.* 1974). With these pre-treatments and additions, acceptable silage can be produced from water hyacinth but palatability is marginal. Baldwin *et al.* (1974) found that of a number of water hyacinth silages prepared and fed to cattle, those with the highest percentages of dried citrus pulp and molasses added were most acceptable to the animals.

Partially dewatered submerged weeds have also been ensiled successfully but in many cases were not palatable enough for use. Water milfoil silage was found to be unpalatable unless the plants were heated prior to being placed in the silo (Sy *et al.* 1973). Heating presumably drove off some of the volatile plant components, which are disagreeable to animals. Bagnall *et al.* (1978) found that hydrilla silage was readily accepted by steers if it was fermented with adequate amounts of dried citrus pulp, ground shelled corn, and propionic acid.

In summary, aquatic plants can be ensiled with proper pretreatment, which may involve chopping, heating, and dewatering. Nutritional quality is generally comparable to more traditional silages, cost is less than dry foods prepared from aquatic weeds but higher than most traditional silages, and palatability is generally very poor without ammendments.

Duckweeds as Animal Feed

Duckweeds are so unique that they must be considered separately from all other aquatic plants. The tiny, free-floating fronds are especially valuable as feed because they do not contain any woody elements. Because of their size and floating habit of growth they require neither mechanical support nor conducting elements. Their reproductive and growth potentials are enormous and they contain more crude protein than most cultivated crops. Table 14.2 compares the productivity and composition of an average sample of duckweed with those of soybeans, cottonseed, peanuts, and alfalfa hay. It can be seen from this table that the amount of protein in the tissue is very high and that the amount of protein that can be harvested from 1 acre in a year's time far exceeds that of even alfalfa. Under some growing conditions, duckweed has been reported to contain as much as 44.7% crude protein (Hillman *et al.* 1978). Furthermore, duckweed is reasonably palatable, even when it is fresh (the common name is derived from the fact that ducks relish it). It does not spoil as readily as most aquatic plants and can be stored for relatively long periods

TABLE 14.2   Comparison of Protein Productivity Between Duckweed and Selected Crop Plants

	Annual production (metric tons/ha)	Crude protein (%)	Relative protein production/ unit area/year
Duckweeds (dry wt)	17.60	37.0	100.0
Soybeans (dry seed)	1.59	41.7	10.2
Cottonseed (dry wt)	0.76	24.9	2.9
Peanuts with skin and hulls (dry wt)	1.59–3.12	23.6	5.7–11.3
Alfalfa hay (sun cured)	4.37–15.70	17.0–15.9	11.4–38.3

Source: Hillman and Culley (1978).

of time after harvesting. One of the major disadvantages is that the nutritional value varies even more widely than most plants with the nature of the water in which the plants were grown.

One highly specialized use of duckweed might be as a replacement for alfalfa meal in poultry rations. Alfalfa is used not only as a source of protein, but as a source of xanthophyll, which is necessary to impart the yellow color to chickens and to egg yolks, which consumers have come to expect. Dried duckweed not only exceeds alfalfa meal in protein content and metabolizable energy, but equals or exceeds it in xanthophyll content as well (Truax *et al.* 1972). At the present time, it is not competitive in cost with alfalfa, but it may someday be grown commercially in effluent from sewage treatment plants where it would serve the dual purpose of removing nutrients from the effluent and providing a source of food for poultry.

Soil Amendments

Relatively few studies have been conducted regarding the use of aquatic weeds as fertilizers or soil conditioners. For the most part, the chemical composition of aquatic plants is suitable for these uses, but the very same factors that interfere with the use of these plants as animal feeds make them impractical as soil amendments. The high water content results in high transportation costs and in anaerobic decomposition with its attendant nitrogen loss. The energy input required to dry the weeds is too great to be practical for a general purpose fertilizer, and the plant nutrient content is too low on a fresh-weight basis to be practical for field application. The variability that occurs with species, location, and stage of growth also presents a problem.

If handled properly, aquatic plants can be composted, and this may prove to be one of their best agricultural uses. Freshly harvested plants are usually too wet and pack too tightly to be composted by themselves, even in high rate composters, which force air through the material being composted (Wile *et al.* 1978). The addition of bulking agents such as peanut shells or ground corn cobs keeps the pile loose and aerobic and allows proper composting to occur. Aquatic plant composts might be useful for certain specialty applications such as making soil-less potting mixes for greenhouse use, where relatively small amounts are required and where slightly higher prices can be tolerated than for general field use.

BIBLIOGRAPHY

AYOADE, G.O., SHARMA, B.M., and SRIDHAR, M.K.C. 1982. Trials of *Pistia stratiodes* L. as animal feed. J. Aquatic Plant Management **20**, 56−57.

BAGNALL, L.O., BALDWIN, J.A., and HENTGES, J.F. 1974. Processing and storage of water hyacinth silage. Hyacinth Control J. **12**, 73−79.

BAGNALL, L.O., DIXON, K.E., and HENTGES, J.F., Jr. 1978. Hydrilla silage production, composition, and acceptability. J. Aquatic Plant Management **16**, 27−31.

BALDWIN, L.O., HENTGES, J.F., Jr., and BAGNALL, L.O. 1974. Preservation and cattle acceptability of water hyacinth silage. Hyacinth Control J. **12**, 79−81.

BOYD, C.E. 1974. Utilization of aquatic plants. *In* Aquatic Vegetation and Its Use and Control, D.S. Mitchell (Editor). United Nations Educational, Scientific, and Cultural Organization, Paris.

CULLEY, D.D., and EPPS, E.A. 1973. Use of duckweed for waste treatment and animal feed. J. Water Pollution Control Federation **45**, 337−347.

GOOSEN, H. 1981. Cape water flowers bloom. South African Panorama **26** (12), 20−23.

HILLMAN, W.S., and CULLEY, D.D., Jr. 1978. The uses of duckweed. American Scientist **66**, 441−451.

KOEGEL, R.G., SY, S.H., BRUHN, H.D., and LIVERMORE, D.F. 1973. Increasing the efficiency of aquatic plant management through processing. J. Aquatic Plant Management **11**, 24-30.

LIANG, J.K., and LOVELL, R.T. 1971. Nutritional value of water hyacinth in channel catfish feeds. Hyacinth Control J. **9**, 40−44.

NATIONAL ACADEMY OF SCIENCES. 1976. Making Aquatic Weeds Useful: Some Perspectives for Developing Countries. Washington, DC.

SCULTHORPE, C.D. 1967. The Biology of Aquatic Vascular Plants. St. Martin's Press, New York.

SY, S.H., KOEGEL, R.G., LIVERMORE, D.F., and BRUHN, H.D. 1973. *Cited by* BAGNALL, L.O., DIXON, K.E. and HENTGES, J.F., Jr. Hydrilla silage production, composition and acceptability. J. Aquatic Plant Management **16**, 27−31.

TRUAX, R.E., CULLEY, D.D., GRIFFITH, M., JOHNSON, W.A. and WOOD, J.P. 1972. Duckweed for chick feed? Louisiana Agriculture **16**, 8−9.

WILE, I., NEIL, J., LUMIS, G., and POS, J. 1978. Production and utilization of aquatic plant compost. J. Aquatic Plant Management **16**, 24−27.

WOLVERTON, B.C., and McKOWAN, M.M. 1976. Water hyacinths for removal of phenols from polluted waters. Aquatic Bot. **2**, 191−201.

Appendix A

Scientific Names of Plants Whose Common Names Appear in the Text

Common Name	Scientific Name
alligator weed	*Althernanthera philoxeroides* (Mart.) Giseb.
arrow arum	*Peltandra virginica* (L.) Kunth.
arrowhead	*Sagittaria* spp.
bladderwort	*Utricularia* spp.
bulrush	*Scirpus* spp.
bur reed	*Sparganium* spp.
cattail	*Typha* spp.
coontail	*Ceratophyllum* spp.
duckweed(s)	Family Lemnaceae
eelgrass	*Zostera marina* L.
elodea	*Elodea* spp.
Eurasian water milfoil	*Myriophyllum spicatum* L.
fragrant water lily	*Nymphaea odorata* Ait.
fanwort	*Cabomba caroliniana* Gray
giant reed	*Phragmites australis* (Cav.) Trin. ex Steud.
hydrilla	*Hydrilla verticillata* (L.f.) Caspary
lotus	*Nelumbo* spp.
naiad	*Najas* spp.
papyrus	*Cyperus papyrus* L.
parrotfeather	*Myriophyllum brasiliense* Camb.
pickerelweed	*Pontederia cordata* L.
pondweed(s)	*Potamogeton* spp.
rice	*Oryza sativa* L.
rushes	*Juncus* spp.
sago pondweed	*Potamogeton pectinatus* L.
sedges	*Carex* spp.
spatterdock	*Nuphar* spp.
spikerush	*Eleocharis* spp.
water bloometjie	*Apanogeton distachyos* L. fil.
water chestnut	*Eleocharis dulcis* (Burm. fil.) Trin. ex Henschel and/or *Trapa natans* L.
watercress	*Rorippa Nastursium-aquaticum* (L.) Hayek; frequently called *Nasturtium officinale* R. Br.
water fern	*Salvinia* spp.
water hyacinth	*Eichornia crassipes* (Mart.) Solms.

Common Name	Scientific Name
water lettuce	*Pistia stratiotes* L.
water lilies	*Nymphaea* spp. and *Nuphar* spp.
water milfoil	*Myriophyllum* spp.
water shield	*Brasenia schreberi* Gmel.
water velvet	*Azolla* spp.
white water lilies	*Nymphaea* spp.
widgeongrass	*Rupia maritima* L.
wild celery	*Vallisneria americana* Michx.
wild rice	*Zizania aquatica* L.

Chemical Names of Herbicides Whose Generic Names Appear in the Text

Generic Name	Chemical Name
amitrole	3-Amino-s-triazole
2,4-D	(2,4-Dichlorophenoxy)acetic acid
dalapon	2,2-Dichloropropionic acid
dichlobenil	2,6-Dichlorobenzonitrile
diquat	6,7-Dihydrodipyrido(1,2-a:2',1'-c) pyrazinediium ion
endothall	7-Oxabicyclo [2.2.1] heptane-2,3-dicarboxylic acid
fenac	2,3,6-Trichlorophenyl acetic acid
silvex	2-(2,4,5-Trichlorophenoxy)propionic acid
simazine	2-Chloro-4,6-bis(ethylamino)-s-triazine

Index